DOMP

Also by R. John Rutten, MD

Black Man, Red Sand

DEEP OCEAN MINING PROJECT

R. JOHN RUTTEN, MD

Edited by Rand J. Rutten

iUniverse, Inc.
Bloomington

DOMP
Deep Ocean Mining Project

Copyright © 2012 by R. John Rutten, MD.

All rights reserved. No part of this book may be used or reproduced by any means, graphic, electronic, or mechanical, including photocopying, recording, taping or by any information storage retrieval system without the written permission of the publisher except in the case of brief quotations embodied in critical articles and reviews.

iUniverse books may be ordered through booksellers or by contacting:

iUniverse
1663 Liberty Drive
Bloomington, IN 47403
www.iuniverse.com
1-800-Authors (1-800-288-4677)

Because of the dynamic nature of the Internet, any web addresses or links contained in this book may have changed since publication and may no longer be valid. The views expressed in this work are solely those of the author and do not necessarily reflect the views of the publisher, and the publisher hereby disclaims them. Some names were changed to protect the identities of the characters.

Any people depicted in stock imagery provided by Thinkstock are models, and such images are being used for illustrative purposes only.
Certain stock imagery © Thinkstock.

ISBN: 978-1-4759-1289-0 (sc)
ISBN: 978-1-4759-1292-0 (hc)
ISBN: 978-1-4759-1291-3 (ebk)

Printed in the United States of America

iUniverse rev. date: 07/26/2012

To Laura Mae Rutten

CONTENTS

Foreword..ix
Introduction...xv

Chapter I	The Cover Story..................1	
Chapter II	Harvey and the Construction of the *HGE*........................9	
Chapter III	The Layout of the Ship and Crews..14	
Chapter IV	Indoctrination..................21	
Chapter V	Soviet Company and the Capture of the TO.......................26	
Chapter VI	Our B Crew's Departure..........33	
Chapter VII	A Safe House on Maui............42	
Chapter VIII	Sick Bay Aboard Ship............52	
Chapter IX	Exploitation of the TO..........60	
Chapter X	Back to Maui for Supplies.......70	
Chapter XI	Heading Out for Disposal of the TO........................78	
Chapter XII	Burial at Sea Ceremony..........84	
Chapter XIII	Getting Ready to Head Home......93	
Chapter XIV	The Long Journey Home..........105	
Chapter XV	R&R and Then Back to the Ship..112	
Chapter XVI	Gearing Up for Another Mission..117	
Chapter XVII	Stretching Out the Legs in Rough Seas.....................124	
Chapter XVIII	Trouble with the Mating Exercise......................138	
Chapter XIX	Returning to Port..............146	
Chapter XX	One Door Closes and Another Opens.........................152	
Chapter XXI	One Last Voyage................168	
Chapter XXII	Debrief and Walking through the Open Door..................183	

```
Afterword.................................193
Glossary of Terms.........................195
Bibliography..............................197
```

FOREWORD

My father first wrote this manuscript upon his return from the Deep Ocean Mining Project (DOMP) from late 1974 to late 1975. He submitted the manuscript for approval to what I understood was the Central Intelligence Agency's Manuscript Review. He was told that the subject matter was classified and that he could not publish his account of the events. My father shelved the *DOMP* manuscript, and it wound up in our family's storage unit back in Santa Barbara, California. In the middle 1980s, Dad was interviewed about the adventure in San Diego, California, and the piece about his role and observations on the still top-secret project was televised on the History Channel.

My Mom and Dad went into government service to Tehran, Iran, for two years, 1976–1978, and then to Panama City, Panama, from 1978–1980. They then went to Alice Springs, Australia, from late 1980 to late 1982, where Dad served as a physician with the United States government personnel stationed there and their dependents. There he volunteered to be a member of the Royal Flying Doctor Service of Australia. He would fly out once or twice a month to hold clinics for Aboriginal tribes in that area. His first book, *Black Man, Red Sand*, which was based on these experiences, was published by Vantage Press Inc. in 1991. From Australia, Mom and Dad went on to complete their government service while stationed in Colon, Panama, from approximately 1983–1985 and then Kinshasa, Zaire, from 1986–1988.

In 2007 my mother, Laura Mae Rutten, was diagnosed with Alzheimer's disease. Dad had survived kidney stones, triple-bypass heart surgery, and bladder cancer, but my mother's caretaking and rapidly advancing dementia quickly took a toll on his health as well. My brother Raul and I moved our parents into separate assisted living facilities, and my brother began managing their affairs.

One day in late 2007, while visiting my parents from my home in El Centro, California, I ran across my Dad's *DOMP* manuscript and showed it to him. He was immediately excited about the thought of trying to get it published again. I took it upon myself to check into resubmitting it for security clearance. In August 2008, I was astounded to learn that nearly half of his manuscript was still classified and could not be published. For the next year, I attempted to write around the classified portions and redactions but kept feeling that the manuscript had lost much of its integrity.

My father passed away on my mother's birthday on January 6, 2009. I had promised him two months before he passed that the manuscript would be published. He asked that I dedicate it to our mother. I learned that in February 2010, most of Project Azorian was declassified. Herewith is the cleared manuscript of Dad's amazing adventure. I must thank Dorothy Grimm, one of my dad's friends in Alice Springs, for all her help with editing this manuscript and for her encouragement in getting it published. I also want to thank Dad's good friend, George Benko, for his assistance in getting permission to publish Dad's manuscript.

Sincerely,
Rand J. Rutten

Hughes Glomar Explorer (HGE) From Wikipedia

Hughes Marine Barge (HMB-1) From Wikipedia

Golf II Class Ballistic Submarine like the *K-129* (T.O.) From Wikipedia

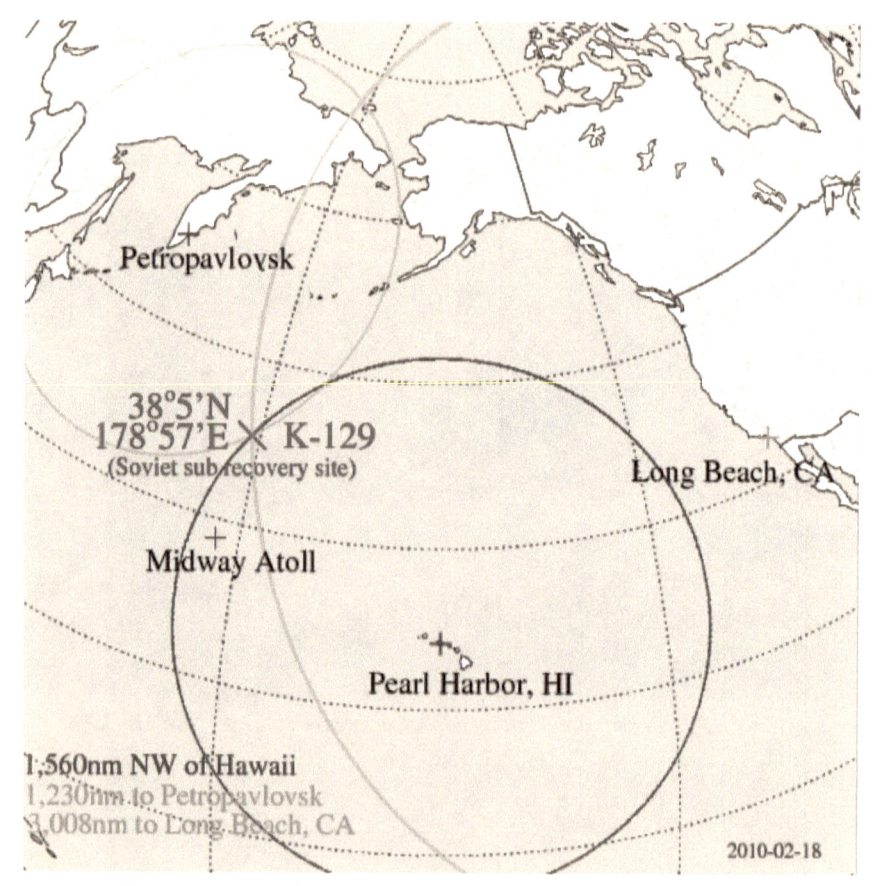

Approximate Location of the Capture of the *K-129* From Wikipedia

INTRODUCTION

This book has developed from my own diaries, letters, and notes during the years 1973 to 1976, and from news articles in various periodicals and magazines published during those years. I have relied on a book written by Wayne Collier, who was employed at the time as a recruiter for Global Marine Inc., in collaboration with Roy Varner, a communications expert, about the subject. The title of their book, copyrighted in 1978, is *A Matter of Risk*. I also drew upon references and accounts provided in Sherry Sontag and Christopher Drew's *Blind Man's Bluff*, published in 1998.

In January 1968, the North Korean Navy attacked and took prisoner the sensitive intelligence-gathering vessel the USS *Pueblo*. US Naval codes, cryptographic devices, and other "top secret" materials were taken captive, along with the surviving crew members.

On March 8, 1968, less than three months later, a Golf II-class Soviet submarine, the *K-129*, went down after a thunderous surface explosion some fifteen hundred miles northwest of Hawaii. Ninety-eight crew members went down with her. There were no survivors. The ship, a diesel-powered vessel carrying nuclear Sark SS-N-5 missiles in vertical launch tubes behind its conning tower and two nuclear-armed torpedoes in its forward launch tubes, was eight years old. The Sarks had a range of some seven hundred nautical miles and contained fifty times the destructive power of the weapons exploded over Nagasaki and Hiroshima that led to the end of World War II in 1945.

The Sarks were targeted for so-called "soft targets," industrial targets in enemy cities, as opposed to "hard targets," armored enemy missile silos, etc. The Soviet ship that went down was 320 feet long, had a cruising speed of seventeen knots, and had a travel range to launch almost anywhere around the world. It was speculated that hydrogen from recharging batteries caused the initial explosion on the *K-129*. There were several more explosions recorded as she sank, and then the final event of implosion of the hull as it plunged into a trench more than three miles below the surface of the Pacific Ocean.

This particular four-year period in history saw a number of submarines from all nations suffering disasters at sea. In May 1968, the USS *Scorpion* went down with ninety-nine seamen aboard, followed by an Israeli vessel, *The Dakar,* that sank in the Mediterranean with sixty-nine crew members aboard. The USS *Thresher* also sank in the mid-Atlantic Ocean in May 1964, and in 1966 the US Air Force lost a hydrogen bomb off Palomares, Spain.

The *Mizar* was a sensitive underwater searching device (like a black-box locater used in an airliner tragedy) designed by the Naval Research Laboratories that was towed methodically behind a surface vessel. It was used in three of the above sinkings with singular success. Based on my supposition, it was also successful in localizing the *K-129* submarine of the Soviet Navy in nearly seventeen thousand feet of water in the Pacific Ocean somewhere off our Hawaiian shores.

At this same time, there were a number of research and development projects underway that involved underwater activities requiring deep-water station keeping. Delco/General Motors Sea Operations (DGMSO) or General Motors (GM) was a company active in this area (with the *Toto* program, to name one) as was Global Marine's vessels The *Cuss I* and the *Glomar Challenger*. In 1968 the *Glomar Challenger* was involved with the *Mohole Project*, a program for drilling into the Mohorovicic discontinuity, or the junction between the earth's crust and its mantle. There was shoptalk of using such vessels for deep ocean mining, harvesting minerals from the ocean floor.

Manganese nodules are a geologic source of concentrated mineral wealth found in rich profusion in many areas of the ocean floor where there has been active volcanic activity. The technology for the "mining" of these mineral sources was hailed to provide a generous return to the prospector. But what privately owned company in the whole world would have the financial capability for such expensive research and development? It would take the resources of a company like Hughes Tool Company. Howard Hughes? Howard Hughes! Summa Corporation would be an excellent partner for Global Marine in this deep-sea secret salvage project.

The Levingston Shipyard in Orange, Texas, had built Global Marine's ships in the past, but at the time they were fully involved in other projects. Thus, with little media fanfare, the keel for the *Hughes Glomar Explorer (HGE)* was laid on December 9, 1971, at the Sun Shipbuilding and Drydock Company in Chester, Pennsylvania, near Philadelphia on the Delaware River. She was designed to be a massive 618 feet long. She could travel at a speed of twelve miles per hour and cost in excess of $350 million to build.

A second vessel would also be critical to the success of the Deep Ocean Mining Project. This would be a barge-like vessel whose ostensible task would be to warehouse the nodules as they were mined. That development would be a combined project of Lockheed Missile and Space Company in Seattle, Washington, and built at the National Steel and Shipbuilding Company in San Diego, California. The barge was to be ten stories high with a net weight of thirty-six thousand tons. One month after the keel was laid for the *HGE*, the *Hughes Marine Barge-1 (HMB-1)* was completed. The entire construction of this massive, unique project took only twenty-two months from start to finish. I am unable to elaborate on the engineering and power capabilities; however, this floating vessel was built to withstand oceanic conditions as fierce as any drilling platform and to lift massive weights from far deeper than any manned submarine could safely travel.

On August 11, 1973, the *HGE* left Philadelphia and headed down the Atlantic coasts of the United States and Central and South America for a voyage of roughly 1,700 miles. She was 115 feet in the beam and so would not fit through the Panama Canal. She refueled in Valparaiso, Chile, just as the coup was underway to depose the regime of Salvador Allende. She arrived in Long Beach, California, within eight minutes of her planned fifty-day transit.

At this time Kennecott, Summa, and Tenneco were the principals involved in serious competitive schemes for deep-ocean mining of manganese nodules.

On January 15, 1974, the *HMB-1* was taken under the Golden Gate Bridge by tugs and towed to Catalina Island, about thirty miles from the coast. Observers speculated at the time as to how many tons of nodules it could hold.

The crew for this expedition was hastily arranged in northern California. It was divided into an A Crew, the "recovery" crew, and the B Crew, the "exploit" crew.

In the spring of 1974, the *HGE* sailed for Catalina Island to meet the *HMB-1*. Built to be housed inside the *HMB-1* was this huge submersible barge with a giant claw mechanism. It was known by the code word Clementine or, as I will refer to it, as the Capture Vehicle (CV). The docking legs that made up the claw were 183 feet high and could be raised and lowered by means of three-foot gears engaging ratchet teeth on vertical sides of the legs. The cover story was that the submersible barge, analogous to the head of a Bissell vacuum cleaner, would "vacuum up" the manganese nodules that were spread over the ocean floor. Practice sessions were performed off the California coast, with the A Crew stringing steel pipe sections from the *HGE* to the submersible barge. The training session was a complete success, and the *HGE* returned to Long Beach to wait deployment. The ideal weather window for the operation was the three months of summer. The *HGE* sailed out into the Pacific Ocean early in June 1974. It arrived on the site over the *K-129*, the target object (TO), on July 5. On arrival, there was no weather window. There were heavy swells in a very hostile ocean. Finally, after five days, the seas abated. The ocean swells were within our working parameters of eighteen feet. Recovery of the downed Soviet submarine was ready to begin.

The first pipe sections were fed through the hole in the ceiling of a large receptacle in the platform of the *HGE* called the moon pool to the bridle of the CV. The connection mechanism to the CV consisted of a single forging of steel 180 feet long, 50 feet wide, and 20 feet high that was said to be the largest unit of steel ever forged. The docking doors of the *HGE* were opened, and the docking legs lowered the CV to 100 feet beneath the hull.

All of a sudden, company! Two Soviet trawlers arrived on site and aggressively monitored the operations. The crew continued with their work despite their apprehensions about the curious, snooping visitors.

A *Maui News* headline read: "Mystery Boat—Rumors Fly!" The story went on to say, "Zipper lipped . . . looked like a band of CIA agents headed for exile." Locals worried that we were mining Hawaii's waters for manganese nodules.

The *HGE* maneuvered into position, and a most extraordinary salvage operation was accomplished. An acquaintance of mine likened the mission at hand to standing on top of the formerly huge 110-story

World Trade Center on a moonless night and lowering fishing line with a 15-foot cage tied to its end down to the sidewalk to pick up a 25-foot steel tube and raise it to the top. Add a thirty-knot wind to simulate the currents that were vectoring the pipe string and one has an analogy to the complexity of this mission.

The TO was located at a depth of more than 16,500 feet. The missile launch tubes were clearly visible above the conning tower of the submarine. One of the three launch tubes had ominously lost its sealed door, apparently in the collision with the ocean bottom. The white nose cone of a missile was threateningly visible.

The most incredible aspects of the harrowing partial recovery of the Soviet *K-129* submarine right under the noses of their homeland comrades were until this year still classified. The portions of the intact submarine that were recovered, including remains of the forsaken Soviet crew, were safely and meticulously analyzed by the B Crew that oversaw this aspect of the project with utmost respect for the dignity of the lost crew members.

The places and events in this story are true. Some names have been changed to protect the privacy of certain individuals who are still alive today. At the time this project began, I was the Director of the Department of Occupational and Preventive Medicine at the Santa Barbara Medical Foundation Clinic in Santa Barbara and had been practicing medicine in the community since 1955. From 1962 to 1976, I was medical consultant to DGMSO in Goleta, a suburban community west of Santa Barbara, and on the site of both the city's airport and the University of California campus. I had been certified as a diver through the Scripps Institute's National Association of Underwater Instructors (NAUI) Program and was a member of the Undersea Medical Society (UMS).

Through the UMS, I had attended courses in diving medicine for physicians sponsored by the National Oceanic and Atmospheric Administration (NOAA) that certified me for the subspecialty of underwater physiology and diving medicine. When Summa Corporation began the deep ocean mining program, they sought out board-certified specialists. As such, a United States government physician who was actively recruiting a medical director for the Summa Project approached me.

The project became public in February 1975 when the *Los Angeles Times* published a story about "Project Jennifer." However, the true name of the project was not publicly known until 2010 as Project Azorian. Herewith then is my account of an exciting odyssey that has only recently been declassified sufficiently to allow telling.

Chapter I

THE COVER STORY

"SECRET PLAN: HUGHES TO MINE OCEAN FLOOR" said the inch-and-a-half-high headline on the front page of the *Los Angeles Herald Examiner* for Monday, November 20, 1972. There was also a front-page photograph of the huge barge, the *Hughes Marine Barge (HMB-1)* that would hold the mining machine designed to mine mineral nodules from the floor of the oceans around the world. The story told of the secret technology that would allow the *Hughes Glomar Explorer (HGE)*, the 618-foot-long ship built through Global Marine by the wealthy and eccentric Howard Hughes, to mate with the huge mining machine contained in the *HMB-1*.

The mineral nodules were said to contain manganese, copper, nickel, cobalt, and other valuable minerals. They are produced by the volcanic hot vents on the ocean floor and seem to be present in huge amounts in selected areas of the deep ocean. It was estimated by mineralogists that there are 1.6 trillion metric tons of these red potato-shaped spheres in the Pacific Ocean alone. The volcanic vents are forming a half a million more metric tons each year. The nodules form like hailstones, in that there is generally a crystallized core found at the center of the friable metallic crust. Often this nitus is a fossilized shark's tooth.

The newspaper stated that the *HMB-1* compound in Redwood City on the San Francisco Bay was shrouded in secrecy. Intrigue and mystery concerning the project were rampant. It was said that city inspectors must call ahead for permission to enter the compound where the construction project was occurring. Federal government officials had queried Hughes Tool Company about the project but had received no answers to the questions about the project.

The concept of a deep ocean mining program was highly competitive at that time. In addition to Hughes Summa Corporation, there were several other privately financed United States ventures with the same idea. At that time, Kennecott Copper Corporation, International Nickel of Canada, and governmentally financed corporations in Japan, Germany, and France were also going forward with nodule mining programs. How far back does the development of this then-current industrial technology go?

Global Marine seemed to have the longest history of building ships with successful position holding capabilities in the open ocean, with the *Cuss I* and the *Glomar Challenger* being used by the oil industry for drilling wells under the sea. There was even a program proposed in the 1960s for drilling into the Mohorovicic discontinuity, the junction between the earth's crust and the mantle layer, for scientific purposes. Remote-controlled robot devices had been designed and constructed to explore the bottom of the ocean at great depths.

The United States Naval Ship *Mizar* towed a Naval Research Laboratories detector in successfully finding underwater metallic objects. In 1968, *Mizar* found the navy's submarine, USS *Scorpion*, in the Atlantic Ocean. In 1969, she found Israel's *The Dakar* in the Mediterranean. She'd also found the USS *Thresher* in 1964 and the French Navy's *Minerve* in the Mediterranean as early as 1952.

Transducers and transponders were also developed to assist in object identification and produce echo sounds from remote positions that could be used via computer to facilitate "station keeping" by surface ships. Vessels at sea have utilized code transducers to facilitate positive identification of friend or foe vessels for years. Another technological breakthrough was Sound Surveillance System (SOSUS) that consisted of bottom-mounted hydrophone arrays connected by underwater cables to facilities on shore. The individual arrays were installed primarily on continental slopes and sea mounts at locations optimized for undistorted long-range acoustic propagation.

The *HGE* had such capabilities for the deep ocean mining project. Another US company had engineered a dynamic positioning system that could hold the 36,000-ton ship in a 150-square-foot surface in 17,000 feet of water, 12-foot seas, and 40-knot winds. The capabilities of this ship and its mining machine were carefully guarded secrets of Summa Corporation, a subsidiary of Hughes Tool Company.

I first became aware of the Deep Ocean Mining Project (DOMP) through Delco/General Motors Sea Operations (DGMSO) Director Dave Parker, MD, in early 1972. He had received communication from Don Flickenger, MD, a consultant for Summa Corporation, requesting permission to visit the plant for the purpose of determining if General Motors could supply the facilities and medical consultant for the DOMP project. Dr. Flickenger was encouraged by Dave Parker, MD, to visit the plant and the clinic where I worked. Dr. Parker was convinced that, since we'd been doing considerable work with the US Navy in its several Anti-Submarine Warfare (ASW) programs since 1962, we could accommodate Dr. Flickenger's needs.

Dr. Flickenger came to the plant and the clinic on March 6, 1972, and was impressed by our capabilities. Dr. Parker, Leo Bancroft, Director of Sea Operations, and I were impressed by the enormous scope of the DOMP as presented by Dr. Flickenger. Costs were discussed, and Dr. Flickenger offered a compensation program for consideration by Dr. Parker for use of GM's facilities and personnel. The GM complement of certified divers numbered about twenty, all qualified for deep open-ocean diving. Dr. Parker agreed to give the offer his most thorough consideration.

That evening, Dr. Flickenger invited Dr. Parker, Mr. Bancroft, our spouses, and me to the Colony House restaurant in Santa Barbara for dinner. The specialty of the house was beef Wellington with Yorkshire pudding. After an excellent meal and short reviews of our four family histories, we parted company with promises to be in contact the following morning regarding the proposals.

The next day at 0800, Dr. Flickenger appeared at the clinic to discuss the program with Mr. Alex Kahn, administrator of the clinic, for my services and the services of the clinic. I introduced Dr. Flickenger to my staff at Automated Multiphasic Health Testing, an arm of the Department of Occupational and Preventive Medicine (DOPM), and to the capabilities of computerized health testing. He was impressed with the state-of-the-art capability that we had online. That evening, my wife Laura and I entertained Dr. Flickenger at our home for dinner and established a close relationship that lasted through the years following this project.

Dr. Flickenger told us Summa Corporation would most likely reject GM's first offer as over budget, and Dr. Parker promised he'd review the project with his advisors and get back to Summa Corporation with a counter offer. By March 22, no word had been received from GM's headquarters in Warren, Michigan. It seemed that DGMSO's participation in the project was unlikely.

April passed with no further communication between General Motors, Summa Corporation, and Dr. Flickenger. Then Mr. Alex Kahn received a phone call from Dr. Flickenger on May 26, 1972. Summa Corporation was willing to accept the clinic's offer for participation in the program, but the clinic would have to find its own sea operations personnel without the assistance of DGMSO. A contract would be signed with the clinic for my services on a schedule of "pay by the day" or "pay by the thirty-day month." Of course, the required physical examinations on the Summa Corporation personnel (mostly certified divers) would be performed through the clinic laboratories and me and would be communicated to a local company in Goleta, California, called Oceanus Inc., another subsidiary of Summa Corporation. I would still continue as medical consultant for DGMSO, director of the clinic's Comprehensive Health Examination Clinic (CHEC) program, and director of DOPM. With the program established through the clinic and Oceanus, a cadre of divers was recruited for the project. These were well-trained, certified ocean divers with several years experience in the field. All were brought through the clinic for their physical examinations according to a specific protocol developed by Dr. Flickenger to fit the specifications of Summa Corporation.

A select group of divers was approved for the program. Perhaps half a dozen others did not make the program because of various disqualifying problems. Hal Sampel was in charge of the divers' programs at Oceanus. He was a former DGMSO officer and a "buddy" diver with me in the Scripps program class that certified us both for open-ocean diving. Additionally, I had been the family physician for Hal and his family for several years. Our Summa consultant to Oceanus was Vic Ellis. Vic was a classic retired navy chief boson-mate type who held more seamanship in his little finger than I held in my whole body. He was a solid-gold asset to our project at sea. Our secretary was Suzie Walker, efficient and well qualified, with years of experience in handling classified matters.

The summer of 1972 passed with occasional contact by telephone with Dr. Flickenger. We were busy handling physical examinations for Summa Corporation through Oceanus, but my total contacts regarding progress of the project were with Suzie Walker, Hal Sampel, and Vic Ellis at Oceanus.

On Friday February 2, 1973, Dr. Flickenger called to inform me that Summa Corporation would like to have Hal and me visit their offices in the Hughes Building in Inglewood, California, just south of Los Angeles International Airport a week later for an indoctrination program into that part of the project and to become acquainted with the project staff. They were located on the fifth floor of the Hughes Building. It's a highly classified office;

Hal and I would be obliged to obtain a special clearance for admission. Hal was able to provide this for me.

Hal and I flew down to Los Angeles International Airport (LAX) United's Convair 880 at 0705 on the morning of Friday, February 9. Hal had my badge, and a van met us at the ramp. Instead of taking a cab on regular roads around the airport, we motored with flashing blue lights across the tarmac to a commercial garage on the south side.

A pair of antennae-protecting globes on the roof distinctively marked the Hughes Building. Hal escorted me through double security gates operated by armed uniformed guards. An attractive receptionist greeted Hal by name, and Hal introduced me to Jennie as the DOMP medical officer. She buzzed an intercom console, and we were asked to proceed to Mr. Harrison's office.

My orientation into the DOMP Project was mind boggling, with a tour of a display of graphic dioramas, audio visual aids, and detailed history, both written and photographic, of the project. The magnitude was just awesome.

I was also introduced to the classified portions of the project that were "need to know" for me. The project was highly compartmentalized so that there were only a few officers who would know the whole project picture at one time. It was at this time that I learned the true scope of the project and the reason for the compartmentalized secrecy. The agenda for the project was as astounding as any sci-fi author could have dreamed in his wildest imagination!

We broke for a buffet lunch in the all-purpose room. Bread, cold cuts, cheeses, fruits, and soft drinks were offered. Then we returned to the administrative end of Summa Corporation's requirements for our participation in the project. There were insurance application forms to fill out for life and disability while on the job. There were forms for listing one's curriculum vitae and biographic history, as well as a person's civil record involving loyalty, criminal, social, and professional evaluations.

We enjoyed lectures and overhead projector visuals by the people who had seen the *Mizar* films. Electrical motors gimbal-mounted on its four corners powered the platform of the remote vehicle so that it could be turned for direction or azimuth and controlled for speed. We were told that a company had manufactured the cameras to withstand the pressure of the ocean at seventeen thousand feet. The strobe lights used on the vehicle were rheostat controlled for light intensity most appropriate for required illumination dictated by film speed.

The diorama showed a Golf II-Class ballistic missile submarine lying on its starboard side on the sloping floor of the ocean. Missile tubes were plainly seen aft of the conning tower. The aft section, perhaps a third of the vessel, was not seen. There was considerable loose debris from the broken-up submarine on the ocean floor, including pages from a book. The water was clear enough that the text could be easily read! The seminar ended in time for Hal and me to catch our United flight back to Santa Barbara.

We were back home by 1700 hours. DOMP was renamed "Azorian." Our mission was to raise a Soviet nuclear submarine! We were advised that we were now charged with the responsibility of top secret security. We took an oath to not refer to Azorian, even to our family, and to continue in the real world with the cover story of DOMP. So be it!

Chapter II

HARVEY AND THE CONSTRUCTION OF THE *HGE*

We now had to wait for further instructions from Dr. Flickenger whether there would be changes in our roster as the files of our Oceanus crew, processed through the clinic, were reviewed by Summa Corporation. These exams were handed to Suzie Walker at Oceanus at noon on Tuesday February 27, 1973. Hal, Vic, and I had an opportunity to describe the course of the project. Eerily, it was proceeding right on time and date with the Program Detail Schedule (PDS) hanging on the wall beside us. On Friday, March 2, 1973 (my birthday), we saw that the wrap-up of the Oceanus's activities had occurred.

The month of April was filled with my return to the now-mundane activities at the clinic. My work with DGMSO continued with activities at the Santa Cruz Acoustic Range Facility (SCARF) and with the Santa Barbara County Medical Society (SBCMS) and politics for the county, state, and the nation, which I was also quite committed to.

On Friday, May 11, 1973, I drove to the Summa Corporation offices at Hal's request. I was informed that the timetable for the project had a new PDS. Summa Corporation and their secrecy agency had now thoroughly reviewed most of the papers that I'd filled out at the seminar in February, and I signed off on them. This time I was given a tour of both floors of "Harvey," as they referred to it at the office. ("Harvey" is the hero in the movie by the same name of the man-sized disappearing rabbit.)

Entry into the Summa offices was through the door on the fifth floor, but there was a stairway in the rear of the office that led to the sixth floor. A large bookcase, just like in a *Get Smart* episode, concealed the entrance to the sixth-floor office. Unless we were employees of the Summa office, we were strongly encouraged to exit from the sixth floor when leaving for the day.

I was given a tour of the full office—both floors. Planning, logistics, accounting, travel arrangements, and Summa's secure communications with Washington, DC, were carried out in these large technologically advanced offices. I was told that afternoon that the Tishman Building nearby on Century Boulevard just north of LAX contained the Global Marine offices. The orientation occupied most of the day, and I was asked to return the next day for the conclusion of the indoctrination program.

I was excited that Saturday, May 12, as I was expecting to be given my dates of potential duty on the project. Strangely, it wasn't difficult to continue using the cover story of DOMP with my colleagues and our families. It certainly appeared plausible. It hung together so well, in fact, that in the March 1974 issue of *Ocean Industry,* the *HGE* was described on the cover as "embarking on mining venture." A detailed description of the ship's capabilities was provided, along with the in-depth cover story for our project.

The PDS showed me that I would be at sea from November 30 to December 22, 1973, and from January 28 to February 26, 1974. There was more briefing on supplying the medical facilities on the *HGE* with necessary equipment and materials. In the meantime, there was the routine work of the clinic, my participation with General Motors and the SCARF program out off the Channel Islands near Santa Barbara, and my family to keep me occupied.

During this time, Global Marine was having its vessels built by Levingston Shipyard in Orange, Texas, but due to a backlog on other projects, they were unable to handle construction of the *HGE*. Consequently, Sun Shipbuilders and Drydock Company in Chester, Pennsylvania, was chosen to build the massive ship. The keel was laid on December 9, 1971.

National Steel and Shipbuilding Company in San Diego built the *HMB-1*. This part of the project was coordinated with Lockheed Missile and Space in Seattle. The *HMB-1* was completed one month after the *HGE* keel was laid. The mining machine housed in the *HMB-1* was also developed and engineered by Lockheed.

The *HGE* was launched on July 23, 1973, in Chester after only twenty-two months in construction. My counterpart on the East Coast was Dr. Jim Borden, the ship's surgeon. He'd been assigned to the ship for the period from July 12 to August 10, 1973. Other personnel included medical technicians Doug Scott, George Benko, and Michael Redmond. Doug had come on board on June 1 and would remain on board until July 8. George was aboard from June 11 to June 23 and then from July 8 to August 10. Mike was on board from May 21 to June 22 and July 7 to August 10. During this period of time, the medics treated eighty-five patients for mostly minor injuries, colds, and such. The first entry on the medical department log is dated June 12, 1973.

The ship's shakedown cruise, East Coast Trials (ECT), took place between July 24 and August 10. Dr. Borden's report described two major injuries on ECT: a fall that resulted in a skull fracture and neck dislocation with subsequent recovery, and a foot run over by a crane that resulted in major reconstructive surgery. Dr. Borden also noted that the pipe handlers were coming into contact with phosphate eaters in the hydraulic fluid of the lifters on an almost daily basis. He cautioned that over the long haul, such exposure could be hazardous. The seals were redesigned to stop the leaking.

On returning to Chester, the ship was prepared for the transit voyage from Chester to Long Beach via the Straits of Magellan because the *HGE*'s beam of just over 115 feet precluded its passage through the Panama Canal. The itinerary for the transit would be St. George, Bermuda; Straits of Magellan; Valparaiso, Chile (for fuel and fresh produce); and on to Long Beach. The ship departed Chester on August 10 and arrived in Long Beach on September 30 a few hours before its estimated time of arrival.

Dr. William Gorgas was the ship's surgeon during its transit. His log showed 118 patients treated in the ship's hospital dispensary. Only one, a fracture of the radial styloid bone, was significant. While fueling in Valparaiso, there was some apprehension on the part of the ship's crew concerning an ongoing coup d'état in Chile, which saw the assassination of the country's president, Salvador Allende. Fortunately, there was no interference with fueling and supplying the odd-looking, three-derricked vessel.

One of the major innovations in design of the *HGE* was the "module" architecture. The shop was designed to be a special type of "container" ship, with each module similar to the van of an eighteen-wheeler. The modules could be produced as living quarters, electronic laboratories, communication centers, kitchens, etc., and "plugged in" wherever there was a need. No structural changes were required on the ship to completely change a module arrangement almost anywhere in the vessel.

Viewing the ship from dockside, one noted the apparent plethora of large and small cranes. Changes in the module arrangement could easily be made by freeing a module to roll on its track to the gunwale and lifting it with one of the many cranes to another location on the ship or to the dock. Then the new module could be installed rather easily and locked into place. Plumbing, electrical, and communication connections were uniformly placed to fit all modules.

Chapter III

THE LAYOUT OF THE SHIP AND CREWS

In April 1968, while DGMSO was planning for an ASW project for the navy called by the name "Sea Spider" for identification and the acronym "Parka" for the site of the project, a diesel-powered Soviet Golf II-class submarine armed with nuclear torpedoes in its forward tubes and SS-N-5 nuclear missiles in vertical launch tubes aft of the conning tower sailed from Vladivostok to patrol the Pacific Ocean. She was picked up by Sound Surveillance System (SOSUS) and tracked as she headed southeasterly.

She was running submerged through the Sea of Japan and into the Pacific. She was about 750 miles northwest of Hawaii when she surfaced to recharge her batteries. Sonar technicians on the US Navy submarine identified the sounds of the charging of the batteries, and then an explosion.

Subsequently, popping and bending sounds were heard, followed by the sound of another explosion much louder than the first. The US Navy submarine reported that it would seem that there had been an incursion of seawater into the battery room. The subsequent production of hydrogen gas and its explosion caused the first loud noise heard.

The secondary sounds of popping were pipes and compartments submitting to the ever-increasing ambient pressure of the seawater as the undersea boat sank uncontrolled. The implosion of the hull when it penetrated its maximum pressure depth produced the loudest signal. The US Navy submarine reported that the TO was down in over 16,500 feet of water and gave the exact coordinates.

A remote device (previously described in the *Mizar's* activities) was used to locate the doomed vessel and take amazingly accurate, clear photos of the wreckage on the sloping floor of the ocean. These are the photos I described in our initial briefing on the program at Harvey. From these detailed pictures, the plans were drawn for the *HGE*. Engineers got their heads together and dreamed up the technology for lifting 8,500,000 pounds from 16,500 feet of ocean. That dream became the reality of this project.

February, March, and April 1973 were busy times at the clinic for me, not only for the DOMP, but also because of extracurricular activities in medical politics to the California Medical Association, the American Medical Association, and the Searle Medidata Automated Multiphasic Health Testing at the clinic. Our recruitment of divers through Oceanus had gone well, and the training often required Hal and the divers to be present at Harvey.

On Friday May 11, 1973, when Hal and I met at Harvey, we were told that sea operations would probably begin on November 30 and go through to December 22. From January 28 to February 26, Internal Systems Testing (IST) for the DOMP would be conducted at sea. I was beginning to feel at home with this fine group of skilled officers and technicians. I was growing anxious to meet our medical techs, George, Doug, and Mike. Also, we had picked up another physician to cover the activities at Harvey on a daily basis, since we now had such a large contingent of people working in the field for procurement and logistics.

Dr. Del Hines had agreed to come on board, even though he had a full time surgical practice in Tucson, Arizona. He would be a backup ship's surgeon for the trips to sea. A third Medic, Jack Thiel, RN from Chicago, Illinois, was in the final stages of processing and should have been available to us by January 1974.

In addition we had studied and organized as well as possible medical facilities available in Long Beach for use in support of the *HGE* on-board medical facilities in the event of serious accident or illness involving the ship's crew while in port. Dr. Borden had done most of this liaison work.

On Sunday, September 30, 1973, Harvey informed us that *HGE* was now at Pier E, Berth 122, at Long Beach. Although I made numerous trips to Harvey for the next few months, I didn't have plausible reasons to visit Pier E. I remained busy with diving activity for DGMSO at their SCARF site. We were providing signatures for the navy's vessels and assisting the air force in recovering their "smart" weapons launched from the sky and recovered from ninety feet of seawater in the Santa Cruz range.

My first trip to Pier E was on Saturday, December 22, 1973. I'd seen pictures of the vessel before but was overwhelmed by the reality in front of me. The central derrick looked like a classic oil-drilling tower, but it was far more than that. It was flanked fore and aft by two docking legs. Each was 183 feet high and capable of lifting four million pounds. The *HMB-1* was 180 feet long, 60 feet wide, and almost 35 feet high, with crisscross hydraulic power lines, strobe lights, and vertical and horizontal propellers. She was deployed from the central moon pool by these docking legs.

When mining operations were underway, the machine was lowered through the bottom of the ship by the docking legs and a bridle was attached to a pipe string supported by the central derrick. The *HGE* was designed to raise or lower seven-thousand-ton loads at a constant rate of six feet per minute.

The moon pool was the resting place of the mining machine when it was traveling to a mining site. She was suspended from the docking legs by huge, horizontal pins centered at each end of the mining machine, and engaging in a V keyhole-shaped notch in the bottom frame of the docking legs fore and aft. The moon pool was 199 feet long, 74 feet wide, and 65 feet high. The floor of the moon pool was designed for two movable doors, each one hundred feet by seventy-five feet, that would open fore and aft like sliding doors. They were operated by rack-and-pinion drives, too, one forward and one aft. The doors were hollow and could be moved away from the hull by flooding them with seawater. To reel them to the hull when closed, they could be made buoyant by blowing out the water. Double rubber gaskets sealed them watertight. The TO was to be delivered to the moon pool for later exploitation away from the recovery site.

There were two complete bridges on the *HGE*, one forward and one aft. The forward bridge was used when the vessel was underway. The after bridge was used for the on-site mining activities, since it was located just aft of the pipe storage area, the pipe transfer crane, and the pipe transfer boom. The bridge was five decks above the main deck and offered a panoramic view of the basic work area.

The double-chamber decompression tanks were located starboard and port above the main deck and two decks below the aft bridge. The inner chamber was eight feet long and four feet in diameter. The outer chamber was four feet long. They were plumbed for air from a bank of high-pressure cylinders located adjacent to the chambers. Each had its own diesel-powered compressor.

The chambers were plumbed for air and oxygen, with the oxygen being supplied by mask with an automatic overboard purge. Electricity and communication lines were also plumbed into the chambers. The medical officer was responsible for appropriate maintenance and testing of the chambers.

The ship's hospital was located on the main deck beneath the aft bridge. It was comprised of a small office, laboratory, x-ray facility, and a larger surgical theater complete with general anesthesia equipment. A regulation steam autoclave was also provided in the surgical theater. There were three rows of upper and lower bunks providing six beds in the hospital area. Surgical instruments, medicine, linen, and equipment storage were provided in drawers and cabinets in the usual marine architectural style. The design was space-savingly small but technically convenient. It was adequate for most surgical emergencies, routine medical problems, or cardiovascular emergencies.

Just aft of the bridge was a large, circular helicopter pad. I used it for my personal exercise track and found that seventeen laps around the circumference were equal to one mile. The crew's mess hall was located below the hospital. It was a very large, cheerfully decorated space suitable for accommodating up to fifty hungry sailors at a time. There were three kitchens providing meals for breakfast at 0400, lunch at 1200, and dinner at 2000. These were the regular times for meals for the three crews providing twenty-four hour watches each day. The menus varied from day to day but were always generous in portion and bistro in taste.

The ship held accommodations for 124 crew members. The medical officer bunked with the chief mess steward. The bunks were upper and lower and remarkably comfortable. Each crewmember had a double steel locker with shelves and hanger space. The "heads" (navy bathrooms) were shared with a crewmember in the port and starboard berthing units and equipped with stainless steel commode and shower and the other usual accommodations. Generally, the officers were housed in the aft bridge section and the technicians in the forward bridge section.

Movie theaters were located on the main deck in both the fore and aft bridge areas. Feature films were shown in both theaters three times a day to accommodate the crews shifting from their eight-hour watches. The theater in the aft section was the next module aft to my bunk. One became used to John Wayne "cuttin' 'em off at the pass" as time went on, but there was no interruption to my sleep anyhow. Each theater held thirty-five persons at a time, and the cooks even supplied the fresh popcorn to go along with the movie selection.

On Friday, December 28, 1973, I received a call from Dr. Jim Borden. He wondered if it would be convenient if he came to the clinic to become acquainted. I was delighted! I had an opportunity to show him through the clinic and introduce him to our administrator, Mr. Alex Kahn, the laboratory and x-ray staffs, and our own CHEC staff. He told me that he was indeed impressed with our facilities and staff. Jim was a youthful, well-developed, lean athlete who was the epitome of the adventurous spirit. I spent the whole afternoon introducing him to the hometown end of the DOMP Project. We even had time to visit the DGMSO facility, where I introduced him to Dr. Parker. That evening, Jim enjoyed dinner at our home with my wife Laura and me on our outdoor patio overlooking the Channel Islands off Santa Barbara. I learned then that the crews would be separated into an A Crew and a B Crew. The "A Crew" would attend the "recovery phase" of the DOMP and the "B Crew" would handle operations on the "exploit phase." Jim was to handle the routine chores at Harvey, and I was to visit there for the increasingly frequent conferences that were required as the project went forward.

On Sunday, March 24, 1974, I was called to Harvey to meet with Dr. Borden and Dr. Flickenger. The tentative date for the deployment of the A Crew would be June 1, 1974. On Wednesday, March 28, the *HGE* left Long Beach for its heralded Pacific Ocean mining operation. I was told at the meeting that I would be going to the San Francisco Bay area for special training in approximately six weeks. According to the PDS schedule, I would be doing my sea leg on this project with the B Crew in August 1974.

Chapter IV

INDOCTRINATION

On Wednesday, May 8, I flew to San Francisco and was driven to an Oceanus office some distance away. I was booked into the Howard Johnson Inn, where Hal Sampel contacted me by telephone. He was to be my escort at the training compound just one hour after my arrival and check-in. He advised me that there would be two days of intensive instruction and training in the activities we would be expected to perform at sea on the project.

Almost as though he had been standing outside the hotel for the last fifty-nine minutes, Hal appeared exactly one hour after he had hung up. We enjoyed the usual amenities on the ten-minute drive from the hotel to the *HMB-1* compound. I was again overwhelmed by the size of this huge floating barge. At about three hundred feet by one hundred feet and over a hundred feet in height, it was designed to hold the CV. It was actually a submersible dry dock that could be moved from place to place by tugs. Armed guards demanded our identification as we drove to the entrance of the tightly guarded compound. Hal displayed the necessary identification, and we passed through the rigid security.

Inside the compound was a large gymnasium-style building. Altogether, there were about forty-five of us who were to be indoctrinated into deep-water salvage, the Russian language, and the Cyrillic alphabet. Basic nuclear physics was also a part of the project training. We were shown a mock-up of the forward part of the downed submarine that had experienced implosion from exceeding its crush capacity under water. It was decorated with trash that was expected to be the terminal result of a water hammer destroying the inside of the submarine. In several hidden spaces, rotting meat had been placed, giving the mock-up the olfactory stimulation that was to be expected in the salvage of such a vessel. There was also a fine powder scattered in several areas that was designed to fluoresce under ultraviolet light.

The first class lasted until 2100 hours, when we were released to return to our hotels by secured vans. We were picked up the following morning at our hotel to continue with our intensive instruction. The van came with six students at a tad after 0600 and picked up twelve of us from our hotel. We stopped along the way to pick up the rest of our forty-five trainees at two other hotels en route. Some of us had managed breakfast of one sort or another, but there was a nice buffet available in the classroom section of the gymnasium. Our first lesson was in Cyrillic instruction. We then proceeded to do a full suit-up and decontamination procedure after we arrived at the mock-up site.

There, radiation outfits were demonstrated. The trainees were each required to don the equipment to become acquainted with the procedure and then were asked to move thorough the mock-up, avoiding those areas with the fluorescing white powder. There were two layers of underwear, the second layer being cotton long johns. The next layer was regular denim shirts and pants. Steel-toed boots were supplied to each of the trainees. We wore surgical gloves with standard work gloves over them. Finally, a paper-like outer layer went over the denims and was duct-taped to the tops of our boots and gloves. A self-contained breathing backpack, a hard hat, and a white hood completed our outfits. We were advised that on-site intercommunication lines would be hung from the ceiling of the moon pool to facilitate conversation with Russian translators and salvage experts. As we passed through the mock-up, we were expected to spell any Cyrillic alphabet words to the attending monitors and describe unusual objects and their precise locations for later identification and photography.

It was indeed a thorough indoctrination. Hal and I flew back to Santa Barbara at 1945 the following afternoon. Now I felt that I was part of an extremely important mission that would have long-range value to national security.

Sunday, June 2, was an important meeting for the medical crew at the Airport Marina Hotel in Los Angeles. Dr. Jim Borden, Dr. Flickenger, Doug Scott, and Jack Thiel were there. Michael Redmond was no longer in the project. Dr. Del Hines was at Harvey and was not at this meeting, where assignments were made. The medical technicians were all going to be on the A Crew for recovery. Dr. Borden would be the medical officer for that sea leg. George Benko would be staying on the ship through both the A Crew and the B Crew activities after enjoying a two-week holiday in Hawaii on return from an A Crew segment. There would be other medical technicians recruited after the *HGE* returned to the exploit site.

At Dr. Borden's suggestion, a tentative recruit had been approached at the clinic, Dr. Jack Lang. Dr. Lang was involved with the intricacies of the emerging technology of open-heart surgery, flew a high performance airplane, and enjoyed scuba diving. Jim felt that the A Crew should have a backup because of the possibility of being "on station" during recovery for an extended amount of time with no possibility of air evacuation of serious cases. It was the consensus that the requirement for major surgical/orthopedic emergency procedures was more likely to happen than medical emergencies. It was agreed to approach Dr. Lang again regarding participation in the project.

That evening the group of us had dinner at the Pieces of Eight restaurant and felt a very strong team bond. Global Marine was accepted as being our prime contractor, with Summa Corporation being the liaison link with Hughes Tool Company.

The next morning, Dr. Lang arrived at LAX, and he went with us to the Hughes Building. Jack received the entire briefing of lectures, dioramas, and explanations. He filled out the same papers we'd filled out and took the same oath of secrecy.

Then he was taken to the *HGE*. After the tour, he acknowledged to me that he too was overwhelmed with the magnitude of the equipment and technology and the incredible task that was planned.

We now had a PDS for the "A Leg" of the project. It included a re-mate off Catalina Island, rig and mining preparation, the integration systems testing, and finally the mining leg. With excitement, we noted that date for on-site commencement of mining would be Thursday, July 4.

On Thursday, June 13, I made a trip to Harvey to sign some extra papers for life insurance. On the chance that the project might be considered an act of war at some point, Summa Corporation covered us with life insurance including that risk. I was also told that *HGE* had departed Long Beach in the night and would be "in the black" until the A Leg was concluded.

Chapter V

SOVIET COMPANY AND THE CAPTURE OF THE TO

On Wednesday, July 10, 1974, I received a telephone call from Dr. Flickenger regarding my compensation for commitment to the B Leg of the project. Generally speaking, it matched what the clinic had been paying me. Dr. Flickenger then informed me that the *HGE* had arrived at the "mining" site as planned on July 4 and was establishing our Automated Station Keeping (ASK) capability. He also told me that there was "company" at the site. The Soviets had intermittent surveillance of this area in the ocean since 1968 and had heard of our deep ocean mining cover project. Two vessels were very close and seemingly expressing a gigantic amount of inquisitiveness.

One was an obvious intelligence-gathering vessel, *The Chasma*, with a satellite navigation dome on its forward bridge and a number of specialized-looking antennae both forward and aft. The other was a smaller oceangoing tugboat that was proudly labeled on its bow with the Cyrillic letters and numbers *Sergei Baikal BS-10*. The crew later nicknamed it "Little Toot." Dr. Flickenger said the crew had gone boldly and openly into their mining program with deployment of the CV through the opened hull of the moon pool.

My discussions with crew members later confirmed everyone's nervousness with the Soviets' proximity, but they felt arrogance and certainty of purpose would better serve the situation than attempting clandestine cover. We had also been informed that there were several female crew members on both vessels.

The pipe-handling team had quickly picked up its pace, thanks to the IST practice off Catalina Island. The *HGE* was designed to handle forty thousand pounds (sixty feet) of pipe every three minutes. The initial rate was more like six feet per second. But, of course, that was subject to stops for maintenance and other minor glitches.

On Thursday, July 11, Hal Sampel telephoned to tell me that the crew wanted to reserve my services for forty days in early August. Apparently all was going well under the critical observation of the two Soviet vessels. I told Laura that we likely would not hear much else about my departure in August unless there were a problem. As they say, "no news is good news."

Six days later, Dr. Flickenger phoned to tell me that one of the crew engineers, John Mackel, had suffered a myocardial infarction on the ship and that Dr. Borden would like my advice on medical evacuation. It was my contention that with the expertly trained staff we had in the medical department and the equipment and medications available, we would not need the heroic measures of medical evacuation or abandonment of the project. Dr. Flickenger agreed that Dr. Borden would provide the crewman quality care.

The next day, Dr. Flickenger telephoned again to say that the medical department at Summa Corporation had agreed with me, and there would be no attempt to take Mr. Mackel off the ship or to interrupt the program to lift the sunken submarine. He also told me that Oceanus had forwarded a purchase order for my leg of the project. It was very exciting to me that the project was truly going forward. I was told that I wouldn't know until three days in advance that I would be required to travel, should the recovery be successful.

Hal Sampel telephoned me on Tuesday, July 23, and said they'd been having storms that produced sea states that required putting action of the lift system plan on hold. The two-axis gimbal system on the *HGE* would provide a stable platform if the seas did not cause a heave of plus/minus of seven and one half feet, a roll of plus/minus of eight and one half degrees, and pitch of plus/minus of five degrees. The heave compensator, I was told, had a stroke of fifteen feet. Hal was saying that the ocean turbulence exceeded those values at the present time. I wagered in my mind that the medical department was seeing a few crew members and dispensing a lot of Dramamine. I imagined the extreme motions there must have been then at the attachment level of the pipe transfer boom on the *HGE*.

On Tuesday, July 30, Hal Sampel and his wife came by our home to talk. Laura was aware of the intensity of the project but was not aware of what our target was yet. Hal's discussion suggested to her, however, that there was indeed more to this project than manganese nodules. After twenty-four years of marriage, she had learned how to read my mind. After Hal and his wife left, we talked together late into the night. Our conversation relieved Laura's apprehensions of my impending departure and her concerns over my involvement with this project. I consoled her with my role being related to my background as a medical physician with the diving industry rather than in the Cold War secrecy of the project's planned discovery of the state of the Soviet's technological sophistication and the eminent danger and urgency for the safe containment of the nuclear warheads on board the sunken submarine.

The Saturday following my conversation with Hal, I treated a diver for decompression sickness in a pressure chamber located in Oxnard, California. I was able to leave the diver for a few hours to go into the chamber and bring it up to surface. I was confident that he would be fine. I planned to check on him the following morning for any problems. I returned home after a five-hour stint. Laura absorbed my recanting of the day's activity. We both wondered, but didn't say, how difficult it would be to be out of touch for forty days while I was at sea.

It was 5:30 a.m. on Monday, August 5, when the bedside telephone rang and awakened me. It was Dr. Flickenger. He told me, "The crew has the TO in the moon pool. We'll need you on board on the 15. Any problem with that?"

I responded quickly, "Gosh, Don, that's wonderful news. No, there's no problem with that. I imagine Hal will brief me on when to be there."

He said, "Yeah, Jack. Someone you know will meet you Thursday the fifteenth at 1310 hours at the United counter at LAX. They'll have all the tickets and vouchers. Give your social security number to Gina for the crew. Good hunting!"

I hung up the phone. Laura was wide awake beside me.

She'd only heard one side of the conversation. "They got it!" I exclaimed incredulously. "I can't believe it! It seemed like such an impossible mission."

Laura said, "When do you go?"

"He said somebody I would know would meet me at the United ticket counter at LAX on the fifteenth, with all the tickets and vouchers. I'm supposed to give Gina at Harvey my social security number for crew assignment."

That left me only ten days to close out my clinic operations and leave my substitute. This was the week that President Nixon resigned, and Gerald Ford became the thirty-eighth President of the United States. A lot of turmoil and uncertainty was in the air at this time. On Saturday, August 10, Dr. Flickenger called me from Seattle. He was to see me on Thursday at LAX. That Monday, Hal called me. He and Mike Redmond were to meet me at the United terminal the next Thursday. I bought my ticket on the United flight from Santa Barbara, anxiously anticipating my exciting adventure to unfold. On Wednesday, Hal phoned and gave me the number for Gina at the Summa Corporation. I called her with my information as directed by Dr. Flickenger.

I had been away a lot with the DGMSO over the past years, and it was never easy for me to say good-bye to my family when I was going to sea for an undetermined time period to participate in a classified project that precluded informing my stoic spouse of the bottom line of our hopeful success.

I had been involved with other classified projects in the past. There was the project called "Deep Jeep," a manned deep submersible vehicle designed by Will Faxton, innumerable diving expeditions to SCARF off the Channel Islands, Sea Spider with the US Navy and DGMSO, and many other cutting-edge projects. In matter of fact, I had been at sea almost four months out of each year for the last five years! In addition to traveling anywhere in Mexico and the United States promoting automated medical health testing programs, I also was involved with conventions for the American Medical Association (AMA), the California Medical Association (CMA), and the American Academy of Family Physicians (AAFP) over the years. But on those adventures, Laura traveled with me.

This time it would be different for us. I wouldn't be able to give daily status reports of my whereabouts or how the mission was going to Laura and our two sons. I remembered the many excursions Laura and I had taken over the years to new and unseen places with our sons, Randy and Raul. I would have to be without their presence on this mission. This time I was embarking on a project that was as technically complicated and as vital to the security of our nation as the accomplishments of Neil Armstrong and Buzz Aldrin in traveling to the moon in July 1969. I felt honored to have been selected for this project and yet quite apprehensive about my eminent departure in such secrecy and intrigue.

Thursday, August 15, finally came around. My first item on the day's agenda was to stop by GM's administrative office near the Santa Barbara Municipal Airport and check out with my occupational nurse there, Mrs. Betty Gilmore, RN. Betty had organized the staff from personnel to have a good-bye party in our health unit for me. I hadn't expected the impromptu and fond wishes from the group but enjoyed it immensely.

The second agenda item was to turn in my pager over at the clinic. Again, many people there expressed to me their sentimental wishes for a safe journey, which I savored for quite some time. Then I went back home to get my gear and say my good-byes to Laura and the boys.

My flight was at 1300 hours. Now that we were on a military project, we counted time by the twenty-four-hour clock. Laura drove me to the airport at about noon. We shared a soul-to-soul hug before I boarded United Flight #858. From the window seat, I could see Laura standing by the outdoor observation deck. She waved to me, and we blew each other kisses. I saw that she was crying, and all of a sudden my eyes welled up. There was a huge lump in my throat that I couldn't swallow.

The Boeing 737 took off to the east. The typical marine layer persisted to two thousand feet, but once we broke through it, the Santa Ynez Mountains appeared with stark clarity. Descending to LAX, I noted that it wasn't too smoggy. We approached the landing north of the racetrack at Hollywood Park. No horses were running that day. I wondered to myself if that was a good sign or not.

Chapter VI

OUR B CREW'S DEPARTURE

True to what Dr. Flickenger said, he and Mike Redmond were waiting at United's flight gate when I arrived that afternoon. Jack was to be our new RN and was to replace Mike, our third tech who had been on the A Crew and experienced the recovery leg. He was returning home to Oahu, Hawaii. He was just full of stories of exciting events they had experienced on that leg.

He told us of a friendly foreign ship which had rendezvoused with the *HGE* at sea because one of its crew members was having chest pain. He said the other vessel made the transfer by a motorboat. Dr. Borden descended the ladder to the small boat and diagnosed the problem as costocondral syndrome, a separation of an anterior rib from its cartilage. This is not uncommon but is extremely painful. Dr. Borden injected the joint with Lidocaine, and the pain disappeared. He suggested to the others in the boat that they wrap the crewman's chest for the next few days. Dr. Borden was obliged to treat the man in the motorboat because security precautions prohibited taking him on the *HGE*. As payment in kind, the friendly ship delivered two cases of Foster's beer and a case of good Scotch whiskey. Dr. Borden had graciously accepted it and promised it would be added to our store for medicinal spirits.

Mike's great story took us all the way to his car. He sure knew his way around the area. He drove us to the secured parking lot at Harvey. The security guard at the front desk recalled me and told me to go to the insurance wing to fill out some more paperwork before I could pass through. I scurried off as my companions were ushered through.

Dr. Flickenger collected those of us who had arrived that day for a briefing that filled us in on some of the events that took place during the recovery. There were about 150 persons assembled in the auditorium that afternoon. Various members of the different operating systems began a day-by-day review of all the events since the arrival on site on July 4. Unfortunately, after thirty-seven years to the date of this writing, I am still unable to reveal certain aspects of this incredible recovery scenario in detail.

There were many engineering challenges to overcome on this never-before-attempted deep-water salvage. A portion of the TO was lost during the lift phase. There were two incidents where the *HGE* was shaken by the lift stresses, which caused the crew members to all consider abandonment of ship. All of this happened under the very intense scrutiny by the Soviet ships that had menacingly dispatched helicopter surveillance of the *HGE*'s deck. Finally, just when the remainder of the TO neared the surface of the crystal clear waters, the Soviet ships turned tail and chugged away over the horizon, leaving the entire crew to blow a sigh of relief. When the CV docked through the moon pool, a large portion of the TO prevented the heavy steel doors from closing so that the water could be pumped out.

Dr. Flickenger and the experts concluded the briefing at that point to the astonished B Crew members. There was a prolonged silence as we mentally digested what we had just been told. Slowly the audience rose and dispersed, each with his own thoughts. Well, we were soon to be on the *HGE* looking into the moon pool and studying the problems of exploitation from each of our professional points of view.

Dr. Flickenger hosted Jack Thiel and me at dinner at a fancy steak house in Marina del Rey that night. The specialty of the house was Japanese-style beer-fed cattle called Wagyu (culottes) steak. Over a fabulous dinner, Dr. Flickenger reinforced the truths of the briefing of the recovery events. It would be a great story to tell the grandchildren some day. But tomorrow is another day.

I stayed with Jack at the safe house at the airport marina that night. We were to our temporary home by 2130 and in bed an hour later. I was up at 0600 on Friday, August 16. Jack had brought a Playboy centerfold poster with him to decorate the safe house, "for all to see when they come home." It was an impressive poster to say the least. We cleaned up our abode and made our way over to the Sheraton Hotel to have breakfast with Dr. Flickenger. He reassured us that all was on track. He also mentioned that I might be asked to ride the ship back to Long Beach after the mission if Dr. Borden was in a hurry to get back to the mainland. We were also told that our cameras would have to be stored at Harvey until we returned.

I wrote a quick note to Laura with these details and told her that I had purchased two medallions at a gift shop at the Sheraton. One was a St. Raphael, patron saint of physicians, and the other was St. Christopher, patron saint of travelers. I placed them both between two seashells on a leather thong around my neck that my sons had given me for a good luck charm and sincerely hoped that both would work for me.

After breakfast we headed back to the safe house and picked up our gear to head to Harvey. Once there, Dr. Flickenger introduced us to J. P., the director of the Global Marine Deep Ocean Mining Project. He had monstrous responsibilities. He exuded nothing but confidence in our mission. He led the short briefing to the B Crew. We then turned our cameras over to the security folks and were admonished not to keep diaries or journals for security reasons. When we arrived in Hawaii, our cover story was that we had not yet done any mining, only testing.

J. P. announced that our flight was waiting, and we would be in Hawaii that night. We lined up our gear at the Walnut Street side of the Harvey complex and boarded buses. We were taken to the Continental Airlines terminal for boarding of the Global Marine charter flight #2450 to Honolulu, Hawaii.

The wives of some of the A Crew joined us in first class. They were to be sharing a Hawaiian holiday with their loved ones, who had made the astounding TO recovery. Jack and Mike and I occupied adjoining seats in the lounge right in front of the Polynesian Pub. Lucy was our flight attendant, and "grog and popcorn" was on the menu. Lucy was kind of skinny but had an infectious smile. I learned that Mike was retired from military service. He was a gunsmith by trade and also did artistic scrimshaw on whales' teeth. He told me he was raised as a Catholic but now professed to atheism. His moral demeanor suggested he was lying! Jack Thiel was from Chicago. His mother made spaghetti sauce, he told us. He had recently taken his two nephews to Chicago from Los Angeles because they had never been there.

A Global Marine representative came by and gave all three of us cocktail tickets. It certainly made the time our flight was on hold at the ramp go by painlessly. We were held up for almost half an hour looking for about twenty people who seemed to have been directed to the wrong terminal. When they were finally accounted for, we locked the doors and were airborne at 1645 hours.

It was clear and sunny, and the mountains stood out vividly. We passed westbound over Santa Barbara at 1700 hours at fourteen thousand feet. Only a tad over two miles from Laura, and I thought to myself that was actually pretty close! I could even see our home from my window seat. She was not in the pool. Heck. She was a "sans suit" swimmer. I decided to keep the thought in mind until I got home from this project.

Mike had flown from Honolulu to Los Angeles the day before just so that he could be on this flight to brief us. He was to head home when we landed in Honolulu. We then passed over Vandenberg Air Force Base at Lompoc, California. I noted that the marine layer was in there with a vengeance as well.

Another of our crew was also a Hawaiian. He was wearing spectacular puka shell lei. He made a present of it to Lucy. She was impressed. We relaxed and chatted and used our three beverage tickets. Dinner on board was a juicy filet mignon with all the trimmings. The view was a gorgeous blue sea below, cirrus clouds above, and a few rows of cumulous clouds with orange tops on them as the sun slowly descended. It was a very comfortable flight.

We arrived in Honolulu at 1905 hours. We were greeted there by a spectacular red sunset behind those same rows of cumulous clouds. We were taken by "wiki wiki" to an Aloha Airlines Boeing 737. At 1945 hours we were off the ground and headed to Kahului, Maui. Again, we were loaded into buses and preceded to Lahaina. The farmers on Maui were burning the sugar cane fields prior to harvest. The whole area there smelled like a caramel factory. The buses turned off the highway at "Prison Street," and we were taken to the Pioneer Inn. The grounds were illuminated with tiki torches. Our gear was unloaded onto the parking lot. It was not a long walk to the pier at the harbor. A glass-bottom boat, the *Coral See*, was waiting to take us to the *HGE*. There wasn't much conversation among us at this point. I thought each of us was lost in his own thoughts of home and the future mission. As reported by Roy Nickerson of the Lahaina Bureau in the Tuesday, August 20, 1974 (coincidentally, Laura's and my twenty-fourth wedding anniversary), issue of the Maui News, "one hundred fifty fresh new crewmen and personnel boarded the *Coral See* and, without so much as an 'aloha,' climbed aboard the Glomar Explorer. They looked like a band of CIA agents heading for exile."

There was a huge welcome for us on board, and we enjoyed the reunion with our old friends. Dr. Borden was indeed anxious to get off and head home to the mainland. He gave me a summary of our heart attack patient, who had already been bused to the airport for the flight home. Jim had an ambitious itinerary he'd set for himself. First he was going to go diving locally with a couple of government divers. Then he was to fly to Los Angeles and then on to Cabo San Lucas in Mexico for more of the same ocean adventures. Jim told me that he would be back in two weeks to pick up his stuff and then proceed to his practice in Alaska. He stored his gear in the closet in our sleeping quarters and locked it up with a big, keyed padlock. He took the keys with him as he departed. I was left to find an empty closet in the hospital to hang my clothes until he returned. It was inconvenient for me, but Jim had had a tough tour so far. I could handle it for a couple of weeks.

We were given a very rapid briefing. The medical supply inventory showed we were out of telopaque x-ray contrast medium and nupercainal (medication for hemorrhoids). We discussed John Mackel's case and also a cholecystitis (gall bladder) case that Dr. Borden had treated with antibiotics. I was also briefed on the complexity of the closure of the moon pool and the potential for radioactive contamination threat that apparently had now been fixed. Surprisingly, there was not much paperwork for me to do. Jim had taken care of it right up to date.

There was a lot of scuttlebutt about going back for the portion of the submarine that had been lost in the recovery. The consensus for a next mission was "no way!" I learned that my bunkmate's daily schedule had him rising at 0330 hours for his shift at 0400. He had already taken the lower bunk. It had been a very long day, and I was exhausted.

I awakened on board the *HGE* on Saturday, August 17, 1974, in Lahaina. I didn't even remember climbing up into my bunk the night before. I was semi-conscious when Carl got up, because I heard him leave at 0400 hours. I dozed until 0530 hours, and then hunger stimulated the juices and I got up. I made my bed, showered, shaved, dressed, checked out our quarters, and then headed for the mess hall. Just as we'd been told, there were some crew members eating breakfast, some lunch, and some dinner. Friendly, smiling, clean uniformed servers with chef's hats staffed the chow line with very amiable dispositions, which did well to direct me toward my challenging appetite.

After a leisurely breakfast, I returned to my cabin and wrote a letter to Laura. We had been informed that we would have regular mail service from the Lahaina Post Office via the *Coral See*. It was now time to meet my other med techs and become acquainted with them. I was impressed with their credentials and meeting with them in person filled me with great confidence. The hospital, which I've already described, was clean, orderly, and well stocked. George had ordered medications and materials from the Kahului hospital that our inventory records showed were depleted. We had no inpatients, and the log revealed most of our problems were muscular-skeletal or upper respiratory. There were a few crew members that had requested minor surgical procedures, such as removal of warts, ingrown toenails, etc. We would get them scheduled for treatment as time permitted.

Our duty shifts were arranged in eight hours for work, eight hours "available," and eight hours of rest and recreation. If it was a minor medical emergency, the rotation was for George to assist me. The next minor emergency call, however, would be Doug with me. My emergency call, though, was all twenty-four hours. The med techs were to come get me or page me when I was needed. Their quarters were just across the passageway from the hospital. It was a good system.

After lunch, I strolled by the moon pool to see what the "catch" looked like. There were a few dressed-out crew members moving around the hull of the raised vessel and a congenial-looking older man standing at the railing of the moon pool. He and I introduced ourselves and gradually exchanged mainland information of our hometowns and families. He told me that his career had been with the German Navy during World War II. By trade he was an engineer and naval architect. His post during the war was with the U-boat missions for Germany's Director for the North Sea Submarine docking locations. After the war, this gentleman established a family and took training at a diving school, where he learned the art of repairing scuttled ships. He became quite successful and eventually sold his shipbuilding company in Danzig, Germany, and emigrated to Chile. He was an expert in wire manufacture, then in the mining industry, and eventually in undersea mining. He certainly appeared to know what undersea mining vessels should look like. I presumed that he was on board because J. P. had reviewed his extensive file and found merit in his education, experience, and training.

Chapter VII

A SAFE HOUSE ON MAUI

On Sunday, August 18, the *HGE* moved during the night about seven miles south of Lahaina to a position offshore of Olowalu Point. The scenery that morning was spectacular. The bow was facing mostly north. To the east the terrain rose to a ridge some four thousand feet high that separated us from the acclaimed Iao Valley. Southeast of us was the peak of Mt. Haleakala. Almost straight west was the island of Lanai, which was mostly owned outright by the Dole Pineapple Company, I was told. Off the bow in the distance one could see the hazy outline of Molokai Island. Then due south was the US Navy's bombardment target island of Kahoolawe, which was to be our anchoring point until the exploitation leg was completed.

The water was crystal clear and warm, and the air temperature was hot! I was told that we would be having lots of boating visitors. For that reason, and also because we would be disposing of some debris (mostly mud from the deep ocean) overboard, we set a float barrier about a hundred yards from all sides of the *HGE*. We didn't want any opponents of deep-ocean mining carrying on demonstrations or being confrontational with our operation to be any closer than that.

I returned to the hospital office and reviewed some of the records of the A Crew who had been treated during the recovery phase. I found a few notebooks that had been kept by Dr. Borden that were titled with a still classified name. As I leafed through them, I became aware that the title was the code name for sensitive items located when the TO was first inspected after the moon pool was pumped dry. There were also records on the artillery and capacities of the vessel and even a small volume that apparently referred to marine life that had come up with the TO as well. When all of the seawater had been pumped out, these critters had come crawling out of the recovered portions of the vessel's hulk.

I hoped that I would have an opportunity to visit the storage van that held these interesting items. I was told that it was a frozen storage vault prepared in advance to handle just such a contingency. I understood that admission to the van was on a "need to know" basis and that without special circumstance, I would likely be denied access. Well, maybe I wasn't that curious.

I walked back to the moon pool catwalk and looked down at the Soviet vessel. It was amazing how much it looked like the mock-up we'd scrambled over in northern California. I decided I would make an appointment to dress out and go onto the floor of the moon pool. It would be valuable to become acquainted with the equipment and personnel up close in case I was called down into the well for any emergency. I planned to inspect both the decompression chambers and start the compressors to acquaint myself with their operation and to make sure they were clean, equipped, and ready for instant use. Lahaina had sought permission to use the chambers for treatment of diving emergencies in the local and tourist population. I was not sure such permission would be granted, but I thought I'd better be ready to answer the questions command would ask should it come to pass.

At 1600 hours the change of shift was announced with the appropriate boson-mate's whistle and accompanying public address announcement. It was time that I gave serious consideration to cleaning myself up for chow and having a chance at that New York steak.

After a leisurely, delicious meal with new friends George and Doug, I then checked back at the hospital to make sure everything there was operating smoothly. Jack was delighted to be relieved for dinner. I rechecked the manuals for operation of the autoclave, the gas machine, the defibrillator machine, x-ray, and auto analyzer. It all looked like standard equipment available in any quality emergency room on the mainland.

It felt strange, but I felt very tired that night. I decided to skip the movie and late night snack and go up to hit the bunk and do some reading. Without knocking, I opened the door to the stateroom. Carl Atkinson was sitting at his desk preparing menus and orders for supplies for the next day with a can of ice-cold beer in his hand. He jumped to conceal the container. I apologized for not knocking and promised that from now on I'd give him a signal before I entered. Carl relaxed with only a little apprehension and accepted my apology. We had all been told that a cardinal rule of life at sea on this vessel was that there would be absolutely no consumption of alcoholic beverages, with the violators being summarily terminated. I rationalized that the restriction could be circumvented if the alcohol was medically prescribed. But it was embarrassing to us both.

Carl apologized to me and said, "I only have one can at night. It's my tranquilizer. I can do without it."

"Carl, I trust you and appreciate your confidence. I can write you a prescription for a can a day if you'd feel better. Or I'll knock before I come in as I promised. I'll leave this particular hour for you to do your paperwork in peace, if that will help."

"Trust me Jack. It's a can a day."

"Not to worry. I trust you Carl." I followed my plan to ready myself for bed. The accommodations were as convenient and complete as any commercial salesman would search out to fit into his per diem. It wasn't like an Intercontinental Hotel or home though. By the time I came out of the bathroom, Carl had finished his work and left. Apparently, he had taken the empty can with him.

I heard the soundtrack of the movie in the next module, but only faintly, even when Humphrey Bogart started shooting. I resolved to see what the audiometer that we used for testing noise levels in various environments registered in the theater and then hit the sack.

Already it was Sunday, August 18. I arose at 0445 hours, slipped on my bathing suit and running shoes, and headed for the helicopter deck. I walked three laps around the deck clockwise to loosen up and then ran seventeen laps. Then I walked another three to cool down and ran another seventeen laps. It was dark when I started, with about a jillion stars shining above us in the sky. I ran on a black indoor-outdoor carpet track, which varied in width from eighteen to thirty inches. When I finished my calculated five-mile trek, the sun had nearly cleared towering Mt. Haleakala. The carpet now looked a brilliant green color. The forty-five minutes of exercise was very stimulating.

As the first light broke, the cane fields began to send columns of smoke straight up in the balmy, calm air. Towering into the sky, the tops of the huge clouds began to show a pink and then turned to bright orange as they topped the mountain ridge to the east and caught the sun's rays. Then the trade winds at altitude caught the smoke, and it streaked horizontally from the top into a parade of orange flags. I imagined then how I would be able to witness this spectacle each morning for the next few weeks.

The atmosphere in the hospital was very somber when I arrived. I learned about one of the A Crew members who had suffered heart problems and had to be stabilized on board. He had been able to last out his leg of the mission and make it back to Maui. He had then flown home with the rest of the A Crew to LAX and from there went to book a flight to San Francisco. The agent told him "There's a flight at the gate right now. If you hurry, you can catch it." When the crewmember arrived at the tarmac, the ground crew was just removing the portable stairs, but they put them back, and the flight attendant opened the door. As the crewmember ran up the steps, he experienced severe chest pain and died on the plane. He had been less than an hour from reuniting with his family. It was also rumored that one of the A Crew had missed his own flight from Honolulu because he'd stopped to make a quick call to the heart attack victim's wife to tell her when he'd be arriving home. After such a beautiful morning, it was a bitter pill for me to swallow.

I tried again to gain access to the mysterious cold storage van. J. P. did his best, but higher authority vetoed my request. I was told that there was "active exploit" to be done, with cutting and removal the following day. I was told that there were some other problems, so the van would have to be moved into the moon pool to be deemed accessible. My curiosity waned after hearing this. I figured that we would be here for a while, and we may find more material that would be required to be stored in the van at a later time.

The local newspapers were accessible each day and fun to read. They were always speculating on the activities of the *HGE* but were almost always clear off base. On that day there was an article that announced that we would be heading for home in five or six days. We were scheduled to have a supervisors' meeting that night. As director of the medical department, I was to be there. We would be given the straight scuttlebutt on any ship movements then—if there were going to be any.

I had yet to see anyone complaining of old-fashioned seasickness. That wasn't strange, though, since we were anchored in very calm water. I had not seen any signs or symptoms, for that matter, of alcoholic hangover, either. On the other hand, there had been no crew liberty since we'd come on board. The only ones going ashore had been those assigned for acquisition of supplies or for mail call.

The next day I was scheduled to go to Kahului for medical supplies. I heard Carl rise and prepare for his day. I lay quietly in my bunk, choosing not to engage him in conversation unless he initiated. He worked hard at not making excessive noise, and I heard him slowly and quietly latch the door as he left. I had learned a bit about Carl's home life. He was a contented sailor, but his family spent an awesome amount of time home alone. It was 0445 hours when I slipped into my swimsuit and strapped on my running shoes.

Once again the morning light was painting Mt. Haleakala pink. No cane fields were burning on this day. I wondered why that was so. I did my usual warm up and then began to run. I tried to run softly as I neared the forward end of the helicopter pad. I had someone ask me if I could be a tad quieter when I ran, since his room was beneath the pad. I kept that in mind as I completed my trek.

After another huge breakfast, I checked in at the hospital. George had scheduled an appointment for a PhD from Los Angeles who was having some chest pain and bleeding hemorrhoids. I didn't think that they had a common etiology. I listened to his history and to his heart and lungs. No significant abnormalities were noted. George had already run an electrocardiogram. I saw that it was perfectly normal, although it was not a stress test. I decided to draw blood for enzymes and lipids. George then ran it in the auto analyzer. The patient's urine was clear and appeared normal.

I then examined the area of his second complaint. No external hemorrhoids were visible. I examined his prostate and rectum with gloved hand. It was a normal prostate, with no masses and no gross blood on the glove. I tested a small amount of the stool on the glove for occult blood. Negative too. I decided to wait to see what the laboratory results showed. For the time being, I prescribed some suppositories for his anal distress. He was given an appointment to return that afternoon when the lab work was done. My thought was that he really wanted to go home.

The exploitation of the TO began in earnest that day. Just like the program in northern California, it was "identify, record, cut, remove, save, and discard." The giant exhaust vacuums sucked air from the work site through eight-inch flexible ducts and released it downwind of the ship. Several crew members were dressed out in self-contained environmental suits resistant to particle radiation as a precaution.

We were told that salvage had begun on the sleeping compartments of the crew in the TO. One couldn't help but speculate on the thoughts that must have gone through the heads of the doomed crew when the disaster was in progress. Milliseconds would have been the time span to take any abandonment action.

At 1300 hours I received a telephone call in the hospital on the secured line from a Dr. Fargo who wished to discuss the stricken A Crew member who had died while boarding the flight home. Apparently, he had been trying to reach Dr. Borden for certain details from the crewman's medical records, but no one seemed to know exactly where the doctor was. Dr. Fargo needed the information for the consulting physicians at Summa Corporation. I heard in detail about the crewmember's rush to board the flight, the sudden onset of his symptoms, the heroic attempts by the flight crew to provide CPR, and finally the mortician's call to the crewman's wife to tell her of the death. I explained to Dr. Fargo that I would do my best to contact Dr. Borden immediately and locate the medical file.

I picked up my wish list for medical supplies from the Maui Memorial Hospital, and at 1500 hours, a boat resembling a big Boston Whaler with a tarpaulin roof arrived. It was about a twenty-minute ride from there in to the Lahaina pier. J. P. and his associate, Ralph, were waiting for me with transportation to the hospital. A local Hawaiian, Bill, was our driver. Ralph asked me how much I thought the medical materials would cost. I told him that it would be a couple of hundred dollars.

Ralph had Bill drive us to a safe house in Kahului. It was a typical Hawaiian residence near the beach with mostly open window areas with bamboo roller shades. It was a two-bedroom, one-bath arrangement with kitchen and study. Ralph pulled the desk in the study away from the wall and removed a back panel from it. He extracted a black canvas pouch, reached in, and handed me three crisp hundred-dollar bills. I pocketed the money. He replaced the pouch and back panel of the desk, and then slid it back up against the wall.

Bill then delivered me to the Maui Memorial Hospital. I presented my list to the supervisor there. She assigned an assistant to put the package together but informed me that my supplies wouldn't be ready until the following day.

Bill drove me back to the safe house. I told Ralph that I would have to make another trip after they had the order together and that the hospital didn't want money until they had the order all together. I handed him back the three hundred dollars and said, "Is this enough to buy a can of Primo beer?"

Ralph laughed heartily, walked over to the fridge, and handed me a cold one. "On a hot day like this, and the good job you're doing, I think the payment is just right," he joked.

It was a hot day. Primo beer is not the best I had tasted in this world, and I resolved that I wouldn't surrender three hundred dollars again for any can of beer.

Bill drove me back to the pier, and I rode the water taxi with the same skipper, Walter, that I had coming from the *HGE*. It was a very choppy ride back. The trade winds had picked up, but they weren't cooling things off; instead they puffed the big cumulous clouds high up into the sky. I knew from my prior trips to Hawaii that there would be a magnificent double sunset to savor that night.

At dinner the med techs told me that all had gone smoothly that day, with no significant illnesses and no surgeries scheduled for the following day. They reported that morale of the crew seemed to be excellent. Everyone was excited with the progress of the exploitation.

I was told that a supervisors' meeting had been scheduled for 1930 hours. It appeared that some major problems in the exploit had been overcome, and the cutting of the infrastructure was set to begin. The plan was to have a special sandblasting party the following day with limited personnel in the moon pool and all the vacuums directed to removal and filtering of the sandblast material from a clear plastic enclosure that was to be constructed around the TO. I was also informed that I would have to be present with the team for that scenario. The consensus was that, if all went well, we would be ready to move out by the weekend.

Chapter VIII

SICK BAY ABOARD SHIP

I missed seeing the double sunset that night, but the cumulous clouds were still hanging in the balmy sky with only a few stars filling in the holes between thunderheads. Some of the tops of the anvil-shaped clouds were still painted glowing lava-colored orange surrounding Mt. Haleakala.

I took in the movie with a box of buttered popcorn. With what was on my mind for the day—the medical supplies delivery, nearing the end of the exploit mission, and a decontamination event the following day—my attention span for the movie only lasted about a reel. I can't even remember what the movie was now. I left the theater and then gently tapped on our door. Hearing no sound from within, I quietly entered our module. I found Carl sound asleep. I showered "navy style" and hit the sack without my usual reading of bedtime novel. I don't even recall rearranging my pillows. I lay down and was able to put my mind to rest for a great night's sleep.

When I awakened the next morning, Tuesday, August 20, I whispered to Laura, "Happy anniversary, Honey." Today was our twenty-fourth wedding anniversary. It was 0430 hours on my desk clock, and Carl had already gone on shift. I hadn't heard him when he got up and only sensed that he wasn't there in our darkened quarters. I dressed in my running shorts and shoes and headed up to the helipad. The outlines of the surrounding islands appeared ghostly from the available moon and starlight and the pale glow of the sun beginning to rise in the east. Once again, there was no smoke from the cane fields, and I wondered why that was so.

I began my warm-up laps and reminisced about our wedding and the vows that Laura and I had made to each other. Laura had been an orphan in fact after she turned twelve, and I had been a de facto orphan at seventeen, due to the outbreak of World War II. I felt a lonely pang in my heart that morning, as it was the first year we had not spent our anniversary together.

I also noted on my run that there was a surprising amount of activity going on in the moon pool already that morning. There had obviously been a change in plans since the supervisors' meeting on Sunday. I would find out what changed when I had "a need to know."

At breakfast I listened to the crew's conversations. There was a rumor going around that the exploitation crew had found more problems with the TO. If those rumors were true, it was going to be a lot longer than the next weekend before we moved to the new location.

I joined the med techs after breakfast in the hospital. There were the usual signatures on logs, charts, and papers that would be going into the mail. I wrote to Laura about my enthusiasm for the mission we were doing and how I wished that I were with her for coffee and our "chirps" that morning to celebrate our anniversary.

I had two minor surgeries on the schedule, and three sign-offs on crew physical examinations certifying medical qualifications for advancements in ranks. I read in the *Maui News* that they now had reneged on their previous stories about our activities, except for the squelching of a rumor that Howard Hughes was now trying to buy the island of Maui. The author, Roy Nickerson, detailed a story on the front page and wrote about the rumors: "[A]s for the rest, including that he himself may be aboard, nobody's talking."

I also saw in a cartoon in the *Honolulu Advertiser* that morning that showed a caricature of the *HGE* in the background at anchor, and in the foreground was a hairy, sandal-clad, weed-smoking local citizen sitting on a bench in front of a Main Street shop. An obvious female tourist with camera in hand was enthusiastically taking the man's photograph. The caption read, "I am NOT Howard Hughes, madam, I am a Lahaina hippie."

There were a few crew members yet on board who had been on board since June who really felt badly that they had not been allowed shore liberty. Our medic, George, was one of them. I made a point in my mind that I would petition the powers that be for some sort of liberty for these people just for the sake of morale.

The two surgeries involved one pigmented nevus (mole) on a crewman's shoulder. George had prepped and draped the site nicely. I infiltrated the site with xylocaine just around the periphery of the lesion and then excised a wide ellipse of normal skin around the lesion. George had accurately labeled the specimen bottle of 10 percent formaldehyde, which was ready for the tissue sample. I wasn't particularly worried about this lesion. It was quite small in diameter, was uniformly black, and had regular borders. There were no enlarged nodes in either axilla. I opined to myself, but didn't record, that it was most likely benign. I based that on the wide borders that I had seen, but would feel reassured with my optimistic but unqualified diagnosis, just in case the pathologist told us we had a problem.

The second surgery was a small actinic keratosis (sun wart) on an older crewman's scalp. George had this gentleman similarly ready when I came into the room. I injected the skin around the site with xylocaine and used the "electric needle" (hyfrecator) to burn it. I curetted the char and then desiccated the raw, oozing vessels at the base. I then applied a Telfa pad and wrote in his chart that he should be fine.

After lunch I went back to take a look at the moon pool from the top. A dressed-out crew was working with cutting torches and heavy vacuums on a portion of the hull that I still haven't been cleared to name. The crew members' conversations were all on a closed circuit to minimize the chance of any nearby eavesdropping. So the only thing audible was the zapping sounds of the cutting torches and the occasional "clunk" of a tool or portion of the TO hitting the steel deck.

I walked over to the decompression chambers and the diesels. The divers took good care to make sure these were "white glove" clean and organized and contained no flammable material. I checked the wiring on the floor fan. It looked brand new. We sure didn't want a spark to appear inside if we were doing an oxygen decompression schedule. I decided that everything there was in good shape.

I then checked out the diesel compressors. There were two ways to start these engines. The first was with an electric starter, which was Plan A. The second was a hand crank that spun a flywheel. When enough potential energy had been transmitted to the heavy flywheel, the engine cylinders were turned over by release of this potential energy to kinetic energy. This was the Plan B method. I tested both methods, and they worked perfectly for both chambers. We probably wouldn't have to utilize divers too much before we returned to anchor off the shore of Catalina Island for the de-mating phase of the project, but one never knows what unanticipated hurdles may need to be overcome. It's been drilled into my nature as a physician to be prepared for anything.

I went down for dinner and got myself seated for a superb meal. Nearby tables of crew members from the exploit crew discussed the new finds on the TO. Some of the physicists seated next to me speculated on the temperature at the front of the water hammer (the sound caused in a pipe containing water that has had steam passed through it) on implosion. Apparently, it was high enough to melt some metals on the downed submarine.

While getting ready for bed that night, Carl and I talked about the day's events. I still felt wound up and decided to try to watch the movie in the theater module again. It was some Hollywood-type musical with too much singing and dancing to calm me. I just couldn't sit through it, so I headed back to a book I had checked out from the onboard library. It was Commander Lloyd Bucher's recounting of North Korea's seizure of the USS *Pueblo* called *Bucher: My Story*. I thought it was particularly apropos since it was contemporaneous in time with the sinking of the *K-129* that had brought us all on this historic mission. I turned out the nightlight at 2100 hours and slipped into another coma, aided by Carl's peaceful snoring.

As my first full week on board began, I arose at the usual time. I performed my usual exercise, shower, and breakfast routines and then met with Mike Cinto from the engineering staff. Mike provided me with a guided tour that morning of the moon pool and the machinery used in the recovery. We wouldn't need the exploit crew to assist us on our tour unless some emergency arose. Our tour began with the docking legs of the *HGE* and their operation. When one was standing in the well of the moon pool and looking up, the sheer enormity of the machinery overhead was mindboggling. I was prepared to describe the engineering capacities and robotic capabilities of this machinery, but to date, I still have not received such approval.

Mike finished with the tour and asked if I had any questions. It seemed miraculous to me that this vessel had been constructed in only twenty-two months. From what I had seen, I was convinced that I really didn't have enough knowledge about the construction or engineering to ask Mike any intelligent questions, so I said that I didn't have any questions. I thanked him for the informative tour and promised to give him a tour through the hospital, should he ever be so inclined. Mike laughed out loud as we parted company at my gracious invitation.

The hospital staff had completed the sick call for the day and told me there was nothing exceptional that they couldn't handle. I had a couple of appointments for consultations the following day, but no imminent minor surgeries on the agenda. I still hadn't heard back from the

Maui Memorial Hospital that our purchase order had been filled.

The dinner menu showed we had a Chinese option for that night, complete with chopsticks for those who chose to practice the art. The chefs expertly prepared Mandarin, Hunan, and Szechwan selections. For dessert I grabbed an apple and walked topside to the helipad.

I was amazed at how many "lookie-loos" had come out in their own boats to take pictures and direct questions to crew members who were present on deck. In general we'd been directed to discourage conversation with the locals. There were some pretty expensive looking boats cruising just outside the skim barriers that surrounded the ship.

I watched the sun set as I ate my apple. I imagined the cumulous clouds looked like dumplings in rows to the west. I made a point then to wait to see the great double sunset that we had that night. The sun at that time was hiding behind the tops of the thunderheads that had built to great altitude. There was a shadow at first, and then their tops turned brilliant silver. Gradually, as the sun descended, the tops become gray again. Then the sun peeped out from the blue space between sea and sky. This was a brilliant rebirth of the full brightness of the day, followed by the ocean's horizon gradually eating up the red ball of the sun until the last bite of the cherry was consumed. It looked like the sun should've made the ocean sizzle as it sunk out of sight. The finale of this pageant-quality display is the change in hue when the gray thunderhead tops turn to pink, then orange, and then red, as the sun fades further east toward Japan. It's a unique show produced almost every night in the South Seas for its residents.

I checked back in at the hospital. George was sitting at his desk writing a letter to home. We discussed the following day's agenda. I was scheduled to go into the well of the moon pool, dress out, and get an orientation on the work on the TO. I learned that we had finally gotten our supplies from the Maui Memorial Hospital, and that they would be delivered the following day.

It was almost 2100 hours and time to retire to my library selection. The movie's audience was just getting out. I could look forward to not being disturbed by the muffled dialogue of the sound track. I found Carl already in bed and asleep. I didn't read for long that night. I turned out the light and replayed in my mind the great meal, the locals in their boats ogling the ship, and the double sunset before falling asleep.

Chapter IX

EXPLOITATION OF THE TO

The Friday, August 23 *Maui News* reported that we'd be here at least another week. It had already been a week since we arrived in Lahaina. A supervisors' committee meeting was set that would explain the changes in the PDS. In the meantime, I was to be in the moon pool for the day. I had been informed that the medical supplies would be delivered, so I knew there was plenty to do that day.

The cool, pre-dawn air was just waiting to be inhaled up on the helipad. Carl was already up and had gone to the galley. I dressed out and began my workout. I smelled a caramel-like odor from the burning sugar cane as I arrived on the deck. I noticed that the days were getting a little shorter. I tried to run gently, if that's even possible, on my track so as not to disturb the crew that were still sleeping beneath me.

I enjoyed the daydreams that come with jogging. There was quite a bit of humidity that morning, which assisted me in producing a healthy sweat. The shower felt marvelous that morning. I ate a hearty breakfast that included crisp cottage fries, but I skipped the eggs.

I headed over to the hospital and began my duty shift. The pathology report for the crewman with the pigmented nevus we removed on Tuesday came in. It was benign. He was scheduled to come in for a dressing change. I looked forward to giving him the good news. I checked the supplies from the Maui hospital against the invoice and saw that they had sent everything I had requested.

I learned that there had indeed been a change in plans on board. Apparently, there were some hot spots here and there in the TO. My dress-out had been postponed until tomorrow. However, that gave me a chance to catch up on some of my paperwork. Just like in private practice, the reports have to be sent in on a regular basis by the medical director to keep the bureaucracy viable. Doug, George, and Jack had them all in order and typed up. All I had to do was read and sign.

Among the papers in my in basket was a Teletype message received in the ship's communication center. "Los Angeles, California. UPI. Actor Peter Fonda returned to Hollywood today from his 18-day Hawaiian cruise with numerous photographs taken from his yacht, *Satan's Doll*, of the secret mining ship *Hughes Glomar Explorer*. A diving friend of Mr. Fonda has identified one person appearing in a photograph as that of medical doctor John Rutten of Santa Barbara, California. Dr. Rutten is well known in diving circles and also an active pilot. A phone call placed to his home in Santa Barbara produced little information on his whereabouts. It was ascertained only that he was not available. Further invest [sic] revealed that a number of friends and relatives have received many letters and postcards from the Hawaiian area in recent days. A local newspaper carried a small photo saying Dr. Rutten was away on vacation."

This must have been generated the day I was on deck remarking about the lookie-loos in their boats outside the barrier. I find it strange how one's point of view changes the perspective.

The weatherman told us that three tropical storms were moving east to west on the satellite picture. We were to expect showers, thunderstorms, wind, and swells by tomorrow. Looking at the sky from the deck, it looked just like it had yesterday, and the day before, and the day before that.

After lunch I took my library book and my shades topside to lie on the helipad and soak up some of the Hawaiian sun. The duty med tech knew where I was for emergencies. I read about two chapters and then dozed off for a short time.

I never received a telephone call that morning to attend the supervisors' committee meeting. It wasn't until dinner that I found out they had held the meeting, but my presence wasn't requested. That suggested to me that the problem that was discovered wasn't medically related and wouldn't involve me.

After dinner I stopped by the hospital and picked up the audiometer. I decided to take it with me to the movie for a little experiment. A John Wayne movie was set for that night, and I was determined to find out the decibel level of the Duke cutting off the bad guys at the pass.

The movie was exciting and noisy. The audiometer read sixty-seven decibels when the "Marshall cut 'em off at the pass." Our readings in the moon pool were in the nearly constant range of eighty decibels because of the huge vacuums. The highest readings we had were from the forward thruster rooms when we were on Automatic Station Keeping (ASK) on recovery. There were five of these thrusters altogether, three forward and two aft. They each developed 1,750 horsepower and assisted the bow and stern to move sideways even when not underway. Our readings recorded sound levels on board as high as 110 decibels. Even with protection, one can't work more than a few minutes in those ranges.

Carl was sound asleep when I came in from the movie, and he was already gone when I awakened the next morning. I thought, *He's sure a good roomie*. His area was always neat, and he cleaned the bathroom after he showered. I did the same, of course. I sure would be glad when Dr. Borden got back. I looked forward to him removing his gear from my locker so that I could reciprocate Carl's tidiness. It was a pain for me to have to change clothes at the hospital, too. I did keep my running gear in a drawer in the desk so that I could dress out for the helipad each morning.

I had noticed that it was easy to forget what day of the week it was with the PDS just running along on a day-to-day requirement. No TGIF on this August 23 morning. I decided to go to the galley. It had been the highlight of each morning to enjoy a good breakfast and conversation with my peers. There was no distinction in our mess as there had been in the navy; there was no separate mess for officers and crew, for example. We were all in one big dining area. By the exit hatch there was always a large bowl of assorted fresh fruit so that one could have a snack later on that morning. I think that was a daily reminder of the team camaraderie we all felt on this project.

I proceeded to Van #14 on that morning to dress out. Everything came off! T-shirt, shorts, and socks were provided. Next we were issued a snowsuit-like, yellow, one-piece cotton garment with a drawstring around the neck. Then yellow plastic foot covers over the socks. The footgear was finished off with high rubber boots. The tops of the boots were then taped shut against the suit with three-inch-wide yellow duct tape. A pair of plastic gloves on the hands was also taped to the yellow suit. A regular pair of work gloves then went over the plastic gloves. Last were the respirator and the hard hat. Away we went.

I felt like an astronaut heading for an Apollo spaceship at that point. We went out of the dressing room and down three stories by ladder to the moon pool's well. The TO seemed huge when I arrived at the floor of the well. I saw the legs of the CV measured about fifteen feet from the hull. The strong back deck was almost thirty feet more. From the top of the deck, it would have been a disastrous fall if one should slip.

Some of the manganese nodules that had come up with the TO were registering "hot." That explained at least one of the reasons why the exploit leg had had to be extended. All that stuff, and other items deemed sensitive, had to be bagged separately and safely for special disposition.

Amazingly, when we climbed back down the strong back to the moon pool deck to exit, I saw that almost three hours had passed. The detectors identified a hot spot on the sole of my right boot. There was no point in trying to decontaminate the boot. It would go through the ultrasound purifier and then into the bag with the hot manganese nodules for special disposal. The clothing was then removed to the skin and placed in a special bag that was labeled "low-level radiation." We then preceded through a detector much like an airport security device. Security personnel performed another thorough examination with a handheld detector, and I apparently passed. I breathed an audible sigh and thought "I'm clean!"

After I dressed in my own clothes, we were instructed to rig one locker in the van for medical dress-out gear for emergencies. It contained a medical bag with my usual instruments and material sufficient to handle most traumas likely to occur, such as broken arms or ankles if someone had an unfortunate accident.

I made the three-story climb up the steel stairs to the main deck and exited from the moon pool. It was a great learning experience. I had no apprehensions after that if I were to be called to dress out in a live emergency situation.

I headed back over to the hospital. The med techs were finishing up their paperwork and continuing medical education reading. I had a great team to work with there. I was told that we'd had the usual kind of day, mostly handling healthy, well-adjusted men of almost all ages.

Oh, yes! There was a message from Summa Corporation that I'd been authorized to continue as medical officer for the return trip to Catalina Island and Long Beach. That was exciting news. I surmised that if they were thinking of Catalina Island, then they might be thinking of a second expedition. On the other hand, it may be just to de-mate the CV with the *HGE*. But it meant I would get a five-day cruise from Maui to Long Beach.

I was up at 0510 hours on Saturday, August 24. Carl had already vanished, as usual. Apparently, Carl was neither a snorer nor a smoker, for which I was grateful. I dressed out for the helipad and went up on deck. I was greeted by a beautiful red and orange sunrise north of the ten-thousand-foot Mt. Haleakala. The orb on the horizon was almost directly over Kahului. It was cool and calm. I again tried to do my workout quietly.

"Olga" was the tropical storm sailing by three hundred miles south of our location, according to the weatherman. "Ione" was a full-blown hurricane with winds at eighty-five knots following Olga's path. Another tropical storm was east of Ione. Garcia, our meteorologist, said he'd never seen so many storms in a row. There was little swell in the channel, and the weather was spectacularly beautiful that morning. I was told that the TO was dissolving slowly. The word was that everything was going smoothly. I went to the hospital and found all to be in order there as well. There were no surgical procedures scheduled for that day, just a few dressings for the med techs to change. I sat down to write an editorial column for the Santa Barbara County Medical Society's bulletin with the theme of "faith and trust."

Next I checked the hospital's mail. The med techs handled everything appropriately with the dressing changes, and I signed off on a few orders and prescriptions. George finished running a few surgical packs through the autoclave. It made me feel more secure to keep a few available and ready just in case someone should take a nasty fall from the decks.

We'd gotten accustomed to seeing a few officials coming aboard from the mainland who had wrangled an invitation from the program director or someone else high up in either Global Marine or from Summa Corporation to get a personal tour of our project. A few were from Washington, DC, or from Nevada. That day we had three visitors. The hospital was always on the itinerary for their tour, since it was well equipped and, as we modestly would admit, the staff was highly qualified. I was designated to give them the tour on this morning and then was invited to join them for lunch when their inspection ended.

J. P. was the host that day. He answered most of the guests' questions candidly and in considerable detail. There were a few areas of a classified nature in which he fudged or tiptoed around responding, but it was more about describing the method of implementation, not procedure. It was interesting to listen to these seminars given to our guests. Occasionally, I would get questions about the medical activities associated with the project, and I was expected to respond with the same direct and confident approach, so I was always alert and attentive.

I had an opportunity to bring the hospital into action shortly after our lunch that day. At 1400 hours a foreman on the pipe crew came in with a history of vague abdominal pain with onset at about 1200 hours. He was finishing his midnight-to-noon shift and was intending to head for dinner. He got down some delicious ham hocks, cabbage and beans, but began to feel nauseated. He sat quietly for a while, but the feeling didn't go away. He went to the head and vomited up most of his meal. He commented that it still tasted good as it came up. He was beginning to feel nauseated again, and indeed he did appear to be a little green. He excused himself, and we gave him an emesis basin to take with him. Not very much came up on this bout.

When he came back to us, he was feeling much better and had a little color in his cheeks again. I palpated his belly and noted that it was soft and not tender. It was, however, quiet on auscultation. This was a little ominous to me, since one of the signs of ileus, a paralysis of an irritated bowel, is the lack of bowel sounds.

I ordered a urinalysis, a white blood cell count, and a differential count (a test to distinguish between an acute disease and something the patient ate). George ran it for me in just a few minutes. I looked at the sample reports and noted the urine was clear. His white blood cell count was ten thousand per cubic millimeter, which is the upper limit of normal. His differential count, however, showed the polymorph nuclear leucocytes (polys, cells directed to bacterial infection control) were slightly elevated.

Since the crewman had finished a twelve-hour shift of hard work and then eaten a substantial supper, he must have felt well during that period of time. I reasoned that if it turned out to be an infection, I could tell in about an hour by getting another WBC and differential counts to see if they were showing more activity. I prescribed a dose of hydroxazine hydrochloride, an anti-anxiety medication, and waited to see what happened next. I also ordered fifty milligrams of Vistaril intramuscularly and put him into a hospital bed. He was asleep in ten minutes. I surmised that he would wake up if his belly hurt.

I left to inspect the moon pool from the overhead catwalk. We were getting higher radiation counts up by the mining machine, apparently from vaporization of metal by the cutting torches. All personnel in the well were dressed out and equipped with respirators. I hoped for their sake that they used the suits diligently, as they had been warned over and over to do in their training classes.

After dinner I headed back over to the hospital. Our patient awakened at 1800 hours with pain in his right lower abdominal quadrant. His temperature was up to ninety-nine degrees Fahrenheit. Jack, the med tech, had done another WBC and differential. The count showed eighteen thousand WBC, and his polys were 87 percent, which was highly elevated. Based on this state of affairs, I determined that we had two options. One was to prepare for surgery here in the hospital, as we'd do if we were at sea. The other was to med-evacuate him to Maui Memorial Hospital. I opted for the latter and thought about how to locate a water taxi for transport.

Communications and arrangements were quickly made, and I heard that a boat had been sent out. By

1945 hours, the boat had not arrived. Our patient was having intermittent cramping abdominal pain and nausea. I decided to give him a quarter grain of morphine sulfate intramuscularly, which quickly relieved his pain.

The boat finally arrived at 2145 hours. Our patient was able to walk down the gangway himself. I observed that the seas were quite choppy then. We placed a Stokes frame, four blankets, and a pillow into the taxi. Jack Thiel rode with our patient.

According to later reports from Jack, when they arrived at the Lahaina pier, an ambulance was waiting for them. They arrived at Maui Memorial Hospital at 2230 hours. Our patient's WBC was then over twenty thousand per cubic millimeter. The hospital was short-handed for surgical team members, so Jack had to stay and assist the surgeon.

The surgery began at 0200 hours on Sunday, August 25. Jack later told me that our patient had a "red hot" appendix, but that it had not ruptured. I finally toppled into bed at 2315 hours. It had been another satisfying day.

Chapter X

BACK TO MAUI FOR SUPPLIES

I was up the next morning at 0500 hours and got up to the helipad. Once again, I hadn't heard anything when Carl hit the deck. I longed for Dr. Borden's return just to get his gear emptied out of the locker. I wondered if he'd ever gotten back to his home in Alaska, since no one yet had heard from him since his departure.

On deck it was the usual beautiful morning. There was a cane field burning way up on the north end of the island that was sending its column of white smoke into the stratosphere. I surmised that it wouldn't be too long until the wisps of the rising smoke began to turn orange.

The weather people had reported in today's "plan for the day" briefing that there's no storm developing near us at this time. But Ione was churning down south with winds at 150 knots! They further advised that in the seventy-two hour forecast, there was a chance it might come right through here. That would indeed require some contingency planning on all our parts.

I strolled to the hospital to check on the paperwork. The lab was neat as a pin; my in basket had a few documents for signature and a note that said Jack Thiel's water taxi was due to arrive from Maui at 0900 hours. I ventured that Jack would sleep until dinner time, so I would have to contain my curiosity about our evacuated patient's surgery.

I remembered that I had a paper to do for a class I was to give at the Santa Barbara City College's Marine Technology Program upon my return. The class would be on the decompression tables for treatment for repetitive diving for the prevention of the problems that working divers experience. My son, Raul, had been my diving buddy ever since he had qualified with the National Association of Underwater Industries (NAUI) in the eighth grade. Raul was one of my students when I taught the program prior to his graduation and eventual deployment for deep sea diving ventures off shore in Indonesia and in the North Sea. So I was looking forward to my presentation. My best friend, Dr. Vernon Friedell, a practicing urologist at our clinic, was the third member of our dive team back home. Vernon and I had been diving together off of Santa Barbara since 1963 and shared ownership in our beloved Bonanza airplane for many years. I hoped to "get wet" with both of them upon my return for a sport dive at our secret spot for abalone not three miles from my home.

The afternoon of Sunday, August 25, slipped away. I had a light lunch and was getting hungry. The TO was almost dissolved now. The well of the moon pool had an awesome amount of cut-up material produced by the exploitation of the submarine structure.

At dinner that night, George and Doug sat with me, and we discussed the case of our patient last night. He was reported by the hospital as doing well. I thought to myself that he was more comfortable with that appendix in a glass jar of formaldehyde. He probably was getting his appetite back by then.

The main entrée on the menu for dinner was chateaubriand. The choice of sauces to enhance it was generous. I chose an au jus with creamed horseradish ladled onto mine and had the baked potato with sour cream and chives to accompany it. Dessert was a piece of fresh strawberry pie.

After dinner I went up to the helipad to watch the sunset. There was an empty milk container box near the hatch. It made a dandy stage for intensive meditation about my presently evolving professional career. It was to be another double sunset that night. The cumulous clouds had risen by my estimate to over forty thousand feet high. There would be red banners blowing from the tops of the clouds when the molten red globe of the sun was quenched in the ocean.

Mt. Haleakala was still in the sunshine while Kahului was in the dusk, and the city's lights were coming on. I sat there until my sunset scenario was completed and then until the stars begin to appear. It was so peaceful and beautiful. I must have sat there for forty-five minutes absorbing this visual poetry. I went below to see what was on for the movie. It was a B crime movie. I chose to go to bed and read Bucher's book instead.

On Monday, August 26, I was scheduled to go check on our patient at the Maui Memorial Hospital. After my early morning exercise and rather Spartan breakfast of Kellogg's Corn Flakes (I hadn't had it in years), I met with Tom Bleaker at the gangway to catch a water taxi in to the Lahaina pier. It was an enjoyable ride in the relatively calm seas in to the city. Inside the barrier surrounding our ship, the water was almost opaque from our muddy discharge. About a quarter mile outside the floating barrier, one could see the bottom, 115 feet more or less below the surface. I felt a little guilty about the pollution we were discharging back into the ocean but reconciled the feeling with the thought that it was all clean mud and that it would settle out within a week after the ship left.

Tom went off on whatever task he had been assigned to, and Jim Dart drove me to the hospital. Our patient was up and walking. He was still on that insipid liquid diet all hospitals offer the immediate postoperative patient: consommé, tea, coffee, and apple or cranberry juice. We handed him his cleared mail that had arrived since he left the ship. Nothing beats letters from home—even junk mail—for raising the spirits of an invalid far from home. We discussed a little of the social life on going on the *HGE* and then departed.

Jim drove me next to the office of Dr. John Warrens, affiliated with the Maui Medical Group on Pauunene Avenue in Kahului. He was a young fellow, tall and athletic, and an active surfer and scuba diver. He had a very busy practice. He gave me a rundown on the progress of our patient and praised the assistance given by Jack Thiel the night of the surgery. His prognosis for our patient was that he would be able to return to the ship the day after tomorrow and to restricted duty after the weekend. Full duty should be in order in about five weeks.

I thanked him for his and the hospital's competent staffs' professional services for members of our crew. I sensed that he was looking for an invitation to come aboard the ship. I told him that I'd see if we couldn't work that out. Of course, there was really no way he could be granted that privilege and most certainly not just by my request.

On the way back to the Lahaina pier, Jim and I stopped off at the Kahului Sears store to shop a little. My favorite chef on board, Harold Castleon, told me he'd sure like a Primo beer T-shirt. He wanted it for his son, who'd been diagnosed with a lymphoma. He'd given me the size. Since Primo is the favorite beer in Maui, it wasn't difficult to find a good selection. Lahaina is a typical Hawaiian tourist town. The huge banyan tree at the harbor had been a landmark for 150 years and had an incredible shade spread of almost an acre.

Tom and I picked up a passenger and we headed back to the ship. The sea was a cobalt blue accented by the white foam from the breaking waves. In this salubrious climate, it was no wonder that so many of the world's vagrants were drawn here like iron filings to a magnet.

On arrival at the gangway, J. P. met us and announced that Bill Hazelman was in sick bay. He'd like me to hasten there and double check what the med techs had found. He had had the same symptoms and signs before, and the docs had always managed to correct the problem with medication. In the conversation, we discovered that it was J. P.'s birthday. We all congratulated him on the occasion.

George and Jack had made the correct diagnosis. They had done appropriate blood work, urinalysis, and an electrocardiogram. They had him on the defibrillator monitor and lying quietly in bed. History revealed that Bill had indeed had this problem before; additionally, he had had a pulmonary embolism in 1968. His current medications contained no agent for arrhythmia control.

The monitor showed us an abnormal ECG, but with no evidence of myocardial infarction. His blood count and chemistries were essentially normal. His blood pressure was within normal limits, but he had the cardiac rhythm of atrial fibrillation and an irregular pulse rate that could be very disconcerting to the patient. It was also ominous in that it could precipitate embolic phenomena. His heart sounds, except for the arrhythmia, were not remarkable. I had Jack give him two hundred milligrams of quinidine sulfate.

I sat bedside with Bill for thirty minutes to small talk. The arrhythmia continued. I then ordered two hundred milligrams of quinidine sulfate again. At 1930 hours, he converted to sinus rhythm. I decided to hold him in sick bay for the night. Bill had no objection to my order. Due to Bill's position in the project, I opted to stay in the sick bay all night with him. The med techs would hold to their usual duty schedule. I decided that we would keep him on clear fluids through the night. I sure did not want him to become nauseated. Besides, all my clothes were down there already, and I would be quite comfortable there—though without my Bucher book.

At 0530 hours on Tuesday, August 27, Bill was asleep, still in sinus rhythm, and Doug told me there had been no abnormalities on his ECG monitor all night. I headed for the helipad and my exercise routine. It was another beautiful sunrise, but it came just a tad later than when we first got here. I enjoyed my usual breakfast and headed for my quarters to dress out for duty. When I arrived, I heard the sound of trickling water. I looked around and discovered that it was coming from the overhead. It looked like a broken water pipe.

It was dripping on the locker with Dr. Borden's storage. I rang the maintenance office, and they said they'd be right there.

I headed back for the hospital. At 0700 hours I received a message from the communication center that Bill was to be replaced in his position because of the arrhythmia. I decided to keep him ambulatory in sick bay for the rest of the day. If he had no recurrence of the arrhythmia, he would be discharged to travel home.

The maintenance people came by to tell me that the leak in the overhead in our sleeping quarters was fixed. They had put a Band-Aid on it, a metal band that screwed tight around the pipe. They told me the pipe was inferior-quality Japanese pipe, and that they had been popping all over the ship. We were fortunate that no real damage was done to our quarters.

There was more good news on that day: Ione was heading north and wouldn't be a problem for us. There was bad news to go along with the good news, though. Two "barographs" (Soviet surveillance vessels) were back in our area. These were the same two vessels that had come near the site the previous week. We were not sure what they were up to. We also received word that we would sail the following night for area III. The navy had requested that all hull penetration areas from the TO be saved for them. Our patient was required to stay at the Maui Memorial Hospital while we were out at sea.

Lunch was a real surprise that afternoon. We were served champagne to celebrate J. P.'s birthday and his departure from the *HGE* that evening. We each received a Dixie cup of the bubbly and toasted one of the few super humans I had ever met. Bill joined us at lunch. He expressed his gratitude to us for our medical treatment. We were delighted with his great results as well.

Mail call! A packet of letters had accumulated for me since the day I left home. It was so great to hear about the activities going on at home. That much mail at one time was overwhelming. I had intended to ration the letters from Laura out but instead ended up reading all of them at once in an orgy of soulful communication. Thank you, Honey, for all the letters you sent while I was gone.

Chapter XI

HEADING OUT FOR DISPOSAL OF THE TO

I woke up on Wednesday, August 28, and learned when I got into the hospital that Bill had had a restful night and felt fine. He was given a light breakfast and released from his monitor. He would be able to move about the ship, but I advised him against shore leave. We were told that Bob Gleaner would be replacing Bill. He was a navy commander, and he had handled Bill's job when the ship was in transit from Bermuda through the Straits of Magellan to Long Beach. Hal Milton was to be the new security officer, and Fred Grogan arrived with Bob. I wasn't told then what Fred's assignment was. He told me at our first conversation that he had passed a kidney stone on the nonstop flight from Washington, DC. That must have been one butt buster of a trip! I ordered a urinalysis to check for hematuria (blood in the urine). It was negative.

It was going to be one of those kinds of days. I had no more than greeted the new people and checked Fred's urine when Tommy Gordon came in. He was working in the well and had sustained a rather severe burn on his left forearm from the steam-cleaning nozzle for decontaminating the mining machine. His left forearm had about fifteen square centimeters of second-degree burns. I bathed it with phisohex, applied some ointment to the lesion, and placed a Telfa four-by-four inch with cotton roller bandage dressing on it. I would see him after dinner and debride any ruptured vesicles and schedule him for a follow-up visit in the morning. I gave Tommy six half grains of codeine to take every four hours for pain. Next I was informed that there was to be an all-hands meeting that night after dinner in the mess hall.

That night there was a rumor at the meeting that there was the possibility of a boarding party attempt when we headed south to dispose of the cut-up portions of the TO. It was felt to be a highly unlikely scenario, but I was glad to see that they were giving consideration to prevention of that type of surprise. It could not have happened while we were at anchorage, but once we left the twelve-mile international limit, it would be a distinct possibility, especially considering our secret cache.

At the meeting, we learned that we wouldn't be moving that night. Bill Hazelman's illness and the subsequent replacements had set us back a couple of days on the PDS. A new schedule was distributed that gave us further details of "disposal." Our patient would be staying on shore while we carried out this part of the project.

On Thursday, August 29, I woke up at my usual time and headed for the helipad. I thought to myself that I must have gotten the treading lightly so as to not disturb my crewmates down pretty well. I had yet to hear of any complaints to the contrary. My clipboard messages had two pieces of interesting information. One was that we were to be underway at 1300 hours bound for the disposal area. The note further said to "keep the ship's bell." It appeared that someone up high in the ranks was going to have a grand desk ornament. The other message was "prepare for transfer of containment vessel from September 21 to October 11." I wasn't sure then if that meant, "get ready for it" or "do it." Three weeks in the "black" at Pier E seemed a little long to me.

Fred Grogan came in to the hospital to have us check on his kidney stone. He passed a specimen for us. He had gross hematuria. He also had uric acid and calcium oxalate crystals in his urine. I ordered an x-ray of his kidneys, ureters, and bladder, which showed no abnormalities there. But he had an old fracture of his right hip that was of some concern for his continuation on the project. Bob Gleaner told me that we'd keep him on board despite these findings.

I was sunbathing on the helipad when we got underway to our next destination. One did not have to worry about neck-popping acceleration on this vessel. We left by way of the Kealakekua Channel.

From my vantage point, I could see that Kahoolawe had some nice, sandy beaches on its southwest end. I went back down to the hospital at 1500 hours to check on the progress of Fred and Bill. They had nothing more to offer on either of their histories to that point. The patients were asymptomatic, which was great news.

I watched the sunset from the bridge that night. It was another beautiful double sunset as we steamed toward the disposal area. I adjourned to our quarters early to get a decent night's sleep.

At midnight Carl woke me up. He was having a lot of pain in his right popliteal space (behind his knee). He told me that he had had the pain occasionally before when he walked fast. I checked out the pulses in his feet. The left was weak at the posterior tibial and dorsalis pedis arteries. His toes had fair color, but he was having obstruction of his peripheral arteries in his legs. He was going to have a lot of trouble with this in the not-to-distant future and may even need a bypass graft someday. "Carl, there's nothing definitive I can do for you at this time, but I'll go down to the hospital and bring you back some medication to relieve the pain and dilate the artery a little. Are you allergic to any medication?"

"No, Doc. Codeine has helped before. It's probably because I went up and down the ladders to the freezers three or four times today, and it's about a four-deck hike."

"Hang in there, Carl. I'll go get some Empirin #3." The medication did get Carl back to sleep. I lay awake contemplating my position at large. It had felt rewarding.

The ship engines were throbbing southward, but very little motion could be perceived. At about 0420 hours on Friday, August 30, the engines stopped. I decided to get up and dressed quietly. Carl was not yet awake. I went up to the helipad to check on our location. It was cloudy and drizzling when I got to my lookout post. We were about a hundred miles south of the big island, according to word at the bridge. I decided to skip my run in the rain. I stood at the edge of the pad and watched the barrels of cut up TO going overboard one by one.

It was another routine day at the start. There were crystal-clear waters and lots of brown sharks with white-tipped dorsal fins around the ship. I could see blue and silver pilot fish following the sharks like they were glued to them way down deep!

At the hospital we had had one significant illness reported when I arrived. One of the crew has bronchitis and a fever. I prescribed penicillin for him on a ten-day course and then got a throat culture.

After dinner that evening, I noticed that we had acquired two little birds and a pigeon. They looked pretty tired as they perched high on one of the antenna arrays.

There was a Soviet trawler in some distress about ten miles off Honolulu, the radio said. They were asking for permission to come in. Our own radar said we were alone with no other ships in our vicinity at all. We were informed that the ocean bottom is sixteen thousand feet down, but there are a number of seamounts in the area. Some are as high as two thousand feet off the bottom in this circle of deep that is ten miles in diameter. There was also an almost-full moon to add to the otherwise eerily quiet on deck.

On Saturday, August 31, it was not raining when I got up to the helipad. I thought it was a beautiful morning to get some exercise in. It was warm and humid, and the thunderheads were already soaring into the stratosphere.

I learned that we had had a near disaster earlier that morning. A load of about a ton came loose from the crane and fell back into the well. Several electric lines and an airline were caught and destroyed in the fall. Fortunately, no personnel were injured. Debris went over the side all day long. I worked on the preventive medicine speech that I was to give at the clinic upon my return.

At 1100 hours we had to button up the moon pool because there was a Soviet satellite scheduled to come over us. No surveillance had been detected so far. The Soviet trawler up in the Honolulu area didn't seem to have much interest in us. Our radar was clear up to the fifty-mile range. We were told that the disposal was about 75 percent complete. We had strong trade winds all day, but the weather folks said it was just the dying Ione. Our seas were running about seven feet, though, from the force of the spent tropical storm.

The hospital had been notified that the "contamination" was now acting like it was supposed to do. We were finding we couldn't remove it from some of the hand railings.

The codeine had helped Carl the day before, and this day he was without symptoms. I checked his pulses, which were still almost nonexistent. I decided that we would have to get him off for some appropriate studies of his peripheral circulation when we got back to a hospital with appropriate imaging facilities.

There wasn't much else significant about the day except the choices on the menu at the mess hall. The final vote for me was steak for breakfast, lamb chops for lunch, and lobster for dinner. I had hoped that I would maybe lose a few pounds on this project. Instead, I was being fed like I was traveling on a cruise ship with Laura.

Chapter XII

BURIAL AT SEA CEREMONY

The brand new month of September began on Sunday. I started my jogging session that morning but had to stop after two miles because of pain in my left knee. It was also hard to stay on the track in the heavy seas. I decided I would have to find some other way to get my exercise for the next days.

I volunteered for lookout duty that day. We were running a grid pattern to make sure nothing that we had been disposing of was popping up from the bottom. I thought how we had put an awesome amount of material over the side the past couple of days; probably enough to make our own seamount.

The ship's bell from the K-129 was on display in the mess hall that morning. I surmised that it was probably made of bronze. It was difficult to tell, because it looked dirty, like it had been dragged through mud. It wasn't mud, though; it wouldn't come clean in the ultrasound cleaner. It was rather banged up and dented at the top. I estimated the bell to be sixteen inches tall and ten inches in diameter. The clapper was separate and similarly ravaged.

At 1930 hours we received a message that Harold Castleon's son, age fifteen and recipient of the Primo beer T-shirt, had been bitten by a rattlesnake! He was in a Santa Clara hospital, and they planned to do surgery on his leg. The boy lived with his mother and stepfather. I spent about forty-five minutes with Harold reassuring him. He was upset, as any Daddy would be.

Good news arrived early on Labor Day, September 2. Word came in that Harold's son was out of the intensive care unit and doing much better. Then at 1030 hours, I had an urgent call to go to the well. It seemed Don Dingman, a healthy young man in his early twenties, had been on one of the mining machine's tines washing it down when he slipped and fell. It was about twenty feet from the tine to the dock of the well. He apparently caught hold of a hose or line and broke his fall. He was lying on his back when I got to him. Initial examination revealed no obvious fractures. He was bruised, but our good luck continued and, miraculously, he was not seriously hurt.

My next patient was George Benko. He seemed extremely nervous. I thought it was cabin fever getting to him. He had been on the *HGE* since Long Beach. He had been a great med tech so far, but he really needed some R&R. I put in a request to Bob Gleaner for two-weeks leave when we hit Maui again. He ought to be able to have a good time for a while then on the short flight over to visit Honolulu.

At 1900 hours we turned tail and headed for Maui at full speed (ten knots). We would be there by noon the following day.

On Tuesday, September 3, the islands of Hawaii were visible on the horizon when I got on the helipad in the morning. I walked it counterclockwise that morning. I thought, *I wonder if that might be the answer to my sore knee. All this time I've been running clockwise and maybe putting the strain on that outside knee on every lap. Maybe I will try it counterclockwise for a while, and then, if all goes well, alternate direction each day.*

The Pacific Ocean phenomenon of huge cumulous clouds on the horizon around us while there was a clear, blue sky where we were stayed with us as we headed for Maui. It looked like rain was falling on Kahoolawe. The small boat from Lahaina was halfway out to us by the time we reached our anchorage at 1230 hours. The water was almost clear when we put down anchor. I could see the hazy, sandy bottom at one hundred feet from the helipad. Nature can sure clean up a lot of man's messes in a hurry when left alone.

The mail arrived, and I had two large envelopes. One had three letters, and the other had one. I discovered that Laura and I had been feeling the same way about each other over the last ten days. I was glad to hear that the whole family had been well.

The folks from headquarters arrived, as well as our patient from the emergency appendectomy. He looked happy and well. A med tech had also come out to replace good old George Benko. Now he could take that long-deserved holiday wherever he wanted to go on the islands.

Doug told us that Dr. Borden was leaving the project as of October 1. That would leave me as full medical director. I really would have to be available at the de-mating procedure at Catalina Island. In addition, I would have a lot more responsibilities with surgery and orthopedics.

Like a mother hen, the ship began to absorb the chicks that had been on leave or liberty in the islands under her wings. We were scheduled to leave after dark that night for the memorial site of the gathered human remains from the TO. They were to be consigned to the deep, as is appropriate with death at sea. I'd heard rumors about the preparations that had been made for an honorable ceremony for the deceased.

We were underway at 2130 hours. We vanished from the Maui coast and headed through the Au Au Channel between Maui and Lanai, and then the Alaalakeiki Channel between Maui and Kahoolawe. We were heading south in the direction of the Big Island of Hawaii.

On Wednesday, September 4, we learned that the Dow Jones Industrial average was 650.74. I wondered what the Dow Jones Industrial average was in the Soviet Union. Maybe there isn't any equivalent. I guessed that I wasn't curious enough to find out.

The PDS showed us arriving on-site for the funeral ceremony at 1200 hours. Dale Norway was in charge of the project; he'd said he would like for the event to be after sundown. I had seen the mass repository for the victims. It was a reinforced metal box, about eight feet by eight feet. It was painted rose-petal red on the outside. Inside there were three tiered bunks on opposite sides of the square, lined with blue Naugahyde. There were two vents on the same sides as the bunks that would prevent crush. There were two flanged steel doors, hinged on the sides of the box over the bunks. Metal plates screwed into the lids secured them. The plan was that Jim Roger would give the eulogy in Russian.

I saw porpoises surfing in our bow wave and playing who can jump the highest all around the ship. There were fewer waves at this particular point in the great Pacific Ocean than we had in our swimming pool at home.

Carl had the chefs prepare a pot of borscht for the evening meal. Borscht is a beet stew made with fresh herbs. He'd also placed boiled mutton, cabbage, and turnips in there. He served it hot with the option of sour cream. I gave it a try. It was delicious. It was a befitting dedication to the young men we were laying to rest that night.

Just as the double sunset began, the crew gathered on deck. Six of the crew members acted as pallbearers. They were dressed in white with white hard hats. The first litter was draped with a Soviet flag, the star at the head of the litter and the hammer and sickle at the foot. Four pallbearers carried the litter from the module to the vault. They ascended a set of wooden steps to a platform at the tip of the vault. Two pallbearers on a similar set of steps in the vault accepted the litter and took it down to the first bunk. The flag was removed in order to cover the next litter with it, since we had only one Soviet flag.

The remains of six Soviet crewmen were intact. The other remains consisted of body parts that had been placed in khaki bags, zippered shut, and then sealed in plastic. The transfer was repeated a total of six times. Three of the complete remains contained ID tags. Each was named in the ceremony. The last litter was filmed as it was deposited in the vault. The "Star Spangled Banner" was played from a recording in dirge time, and then the Soviet National Anthem was played in a similar cadence.

Dale Norway gave a short English recitation to us about having no knowledge of the Soviet Navy's burial ceremony, but that we would do our best to honor these dead seamen.

He ascribed the responsibility for their deaths to the mutual distrust between the United States and the USSR and finished with the honor of being "one who goes down to the sea in ships." He then translated the eulogy into Russian. He concluded the ceremony with a solemn prayer that they be committed to repose until such time as God recovers all men buried in the sea.

The vault was sealed and lifted from the upper forward deck. The ship was running south. The western sky on the starboard side was red with sunset fading behind a skyline of irregular cumulous clouds. The vault was visible from the surface for perhaps fifty feet in the gloaming. We were about twenty miles from our previous dumping ground.

Blue plastic formed a false wall around the vault. Even the deck was covered with blue plastic. The after vault was draped with an American flag and a Soviet flag, each five by eight feet. It was the same Soviet flag that had been used for the litter shrouds. Everyone on board was deep in thought and somber as the vault was let go into the open ocean.

The next morning, Thursday, September 5, I decided to give my old exercise program a retry. I walked two miles counterclockwise. My left knee was limbering up a little. I also pulled my swimming suit up high during sunbath on the "steel beach." I toasted for forty minutes on each side. In spite of the fantastic weather, I was beginning to look like a marshmallow because I had been inside so much.

The *HGE* was running in circles that morning. We were not supposed to reach Lahaina until 0700 hours tomorrow morning. We were supposed to straighten out the circles and head for our mooring about 1600 hours that afternoon.

We had another near disaster on this day. "Shorty" Rosemont was cleaning the wall of the well with trichlorethylene. He was wearing a respirator, whose head straps made it inconvenient to wear a hard hat. One hundred twenty feet above him another worker on the catwalk dropped a shackle bolt. It struck Shorty on the respirator hose and glanced off his third left costochondral joint. It knocked him down, and he struck the back of his hand on a basket. There was no damage, except for a bruise on his left chest and a scratch on his hand. It missed his unprotected head by only inches.

The conversation at dinner was limited that night. I thought that each of us was reflecting on the ceremony from the night before and finding the appropriate cranny in our memory to store something as momentous as that was. We couldn't share the events with anyone except a shipmate.

I didn't see any effect from the sun when I finished on the steel beach on the morning of Friday, September 6. However, that evening when I took my shower before hitting the sack, I noticed a definite tingling of the skin and a little blush on my legs and chest. I was glad I never gave it any more than I had. I also noticed that the right side of my nose, where I had a little keratosis, was now broken down in the center. It was my first basal cell epithelioma—skin cancer. I resolved that I would have it taken care of as soon as I got back to the clinic. I crawled into my bunk at 2200 hours with my book and quickly fell asleep.

Just as I had thought the day before, we were entering the Kealaikahiki Channel between Kahoolawe and Lanai at 0515 hours. I was up walking my course counterclockwise to lessen the impact on my left knee. That morning I got a lesson on distinguishing Betelgeuse, Rigel, Sirius, and Canopus cloud types from our radioman, Ken Atkins (a pilot, too). He had joined me on the helipad. We agreed that we were at just about the same anchorage as before.

Later that morning I received mail from Laura. She told me that our son, Raul, was vacationing with his friend, Dino Ventura, and Dino's family in Hilo, Hawaii. I thought what a coincidence it was that I was within shouting distance from family on this secret project. Laura gave me his address in Hilo. I decided that any letter I sent to him would have to go through Summa Corporation before going out, and by the time it got to the address, Raul would have headed home.

I also received my service club's (Goleta Lions Club) monthly newsletter, the *Atelog*. Atelog, of course, is Goleta (our home town) spelled backward. I take great pride in coming up with the name and being its editor in chief over the past fifteen years. This cheered me up and made feel at home again briefly.

The rest of the afternoon of Saturday, September 7, was taken care of with reports coming in and going out for routine business. I watched the double sunset from the steel beach again that night and enjoyed an apple for dessert. I couldn't help but have thoughts about a second mission with the project. I was in bed and asleep at 2200 hours.

The next morning, I heard Carl when he got up at 0400 hours. His legs were giving him trouble again. We had taken a port list of four degrees yesterday to clean some starboard areas. It felt strange to be walking uphill so much of the day. Bill Borders, one of our divers, came in looking for some fluorescein dye. It was supposed to go into the saltwater system to track a leak. Bill said they were scheduled to be in the water by 0800 hours tomorrow. We didn't have that much fluorescein dye available, only what was on the little ophthalmic strips to search the cornea of the eye for scratches.

I also heard that Bill Hazelman was in the Tripler Hospital with another episode of irregular heart rhythm. I was given word that we wouldn't be underway again until Monday, September 9, for now.

That would give us the weekend to get supplied with vittles and fuel.

At lunch I overheard the engineers talking about the needed repairs to the CV if called upon for the second mission. The steel in the mining machine was a super type of steel. It was very hard and very brittle, but also very light and very strong. The engineers had a name for it—"marfixing" or "marfacing," something like that. The quantity involved in the repair will represent the whole national production for one year!

I performed the routine Saturday of the sanitary inspection of the kitchen, cold stores, freezers, heads, distillers, and the power plant. Then I finished the nutrition portion of my scientific-notes talk. There wasn't much work in the hospital for the rest of the day.

After dinner I went up to the steel beach for the double sunset, and Jim Roger came by. We watched the planets become visible in the moonless dusk, and then the familiar stars I've come to recognize. I was in bed and asleep by 2200 hours and slept well.

Chapter XIII

GETTING READY TO HEAD HOME

Sunday, September 8, began as the usual Holy Day should for me, but became a little messy as it progressed into the morning. I was now up to three miles walking counterclockwise alternating with clockwise laps. My knee was really much better, and I was hardly aware of any pain or discomfort during my routine.

   The sunrise over Maui was spectacular. Balmy and calm, the ridge to the east changed colors from blush to fuchsia almost by the minute. The only noise was that of the gulls arguing about the morsels in the sea that they were bobbing for. I went back up to our quarters to dress for my work shift and was summoned by my pager.

   One of our crewmen had been in the well cleaning with trichlorethylene when he splashed some over his safety goggles and into his eyes. I couldn't imagine how he had done that, but it did happen on occasion. They always swear that they were wearing their protective gear, but one just has to wonder. It was becoming routine enough that we would just do a copious irrigation and send them back to work.

   Before I could go back to my routine duties, another moon pool crewman fell in the well, striking his right lateral leg on some gears. Examination revealed tenderness but no limitation of motion and no evidence of fracture. Poor guy, it just wasn't his day. The med techs told me later that he came in about an hour after I had gone. He'd dropped some steel on his ankle. They had checked him out like I had done and couldn't find even a bruise. They gave him some oil of wintergreen ointment (a placebo to me) to rub into the tenderness.

   Our next patient was John Brandywine who came in after he had lifted some steel I-beam pieces and developed a right indirect inguinal hernia. It was a recurrence of a previous hernia. We located a truss to provide him with abdominal support until we could send him ashore for a further evaluation the following day.

The next patient was John Walters, who was departing the ship for liberty. He had fallen while transferring from the *HGE*'s accommodation platform (Jupiter) into a small boat. He didn't fall cleanly into the water; instead, he caught the gunwale of the small boat with his hands and pulled himself into the boat. About a mile from the platform he fainted, so they brought him back. Examination revealed extremely tender abrasions and contusions to his right iliac crest. He had no belly or hip pain. His extremities all moved satisfactorily. He had a good femoral pulse at eighty-four per minute on the right side. He told me he felt better; I told him to go ahead to shore. When they reached the dock in Lahaina, John stepped out of the boat and promptly fainted again. By radio, I advised that they take him to Maui Memorial Hospital's emergency room for x-rays of the iliac crest. At 2200 hours I was told he had had x-rays and that an orthopedic consultation confirmed a chip fracture of the right iliac crest—an extremely painful trauma. I was later told he would be confined to the hospital for four or five days. This meant he would miss shipping out with us with the scheduled departure at 0900 hours the following day.

This same afternoon at lunchtime, I'd returned to our bunk to pick up some medical notes. Carl was ordering final provisions for our departure. He commented that he'd been having no problem with his leg over the last couple of days. He'd been off cigarettes for nearly a week. I congratulated him. He offered me a can of Primo, and I accepted. It tasted good!

At dinner I learned that I wasn't going to get copies of the photos of the two Soviet ships we had monitoring us at the start leg of the project. I was told that I wasn't sufficiently cleared to receive them. I got to bed by 2300 hours and hoped I would sleep well for our morning departure.

On Monday, September 9, we "upped" the anchor before we "upped" the landing platform. The bow thrusters turned the ship around on a dime, and we proceeded away from Lahaina and steamed down the Alaalakeiki Channel at 0900 hours. We left a small buoy (Bedford Griggs) behind with a sign that said on one side, "Reserved for the *HGE*." The other side read, "No Mining Without Permit." The natives were sure to have fun with that one.

We sailed past the small volcano cone in the channel. It had a light on the Kahoolawe side; otherwise, it was abandoned. The wind side of Kahoolawe was sharp lava cliffs, and so was the wind side of Maui. The swell was big then, and we hit wind on our bow that cut our speed way down.

The wind side of the eastern end of Maui (Hana) is green and forested above two thousand feet. It looked like the Danish town of Solvang in the Santa Ynez Valley just over the mountains from Goleta. Farms, barns, and silos were scattered across the lush, green carpet surrounding the rising mountains. I could see waterfalls cascade hundreds of feet down verdant lava-based cliffs. At sundown the tip of Maui was still visible low on the horizon from the ship's deck.

Keith Milton was a navy type who came in to the sick bay the next day looking like he was all pumped up with adrenalin. He began rattling off a lot of verbiage about calling a board of inquiry into John Walters' accident. I couldn't help but stare at him in amazement. His eyes were lit up, his juices were flowing, and he was happy as a Cub Scout on his first overnight. I got his adrenalin back down and his eyeballs retracted by declaring that a board of inquiry wouldn't be necessary. I invited him to come in anytime and talk the event over if it would make him feel better. He seemed mollified. He wondered what would come of this in the future.

Buster Coperman, about thirty years old, had been working in the well throughout the exploitation phase. He came to me at the dinner table that night to tell me of his strange pain bilaterally in his flanks. He thought he should be evacuated by helicopter. His exam was completely negative. His attitude made me think he might be starting a play for compensation for radioactive material exposure. That night the theater was rerunning tapes of the capture of the TO. It was very exciting to watch and think that I too was connected to this project.

Tuesday, September 10, was a routine morning for me. Carl hadn't complained of his claudication since that first episode. I didn't think he'd really quit the cigarettes completely, but he'd sure cut down a lot. I hadn't seen or smelled him light up in our quarters since the acute episode. I was back up to three miles now walking. I would alternate my direction every seventeen laps around the helipad. My knee wasn't swelling or hurting at all anymore. Maybe I'd start to jog some of the laps. I think the sleepers below deck were probably delighted that I had been walking. I expected to get some flak from them when I started running again.

It was a quiet day at sea. There were a couple of rain showers and the usual scattered cumulous clouds that liked to hide at the horizon. Jack Thiel had taken on a new project. He was building a scale model of a Flying Tiger's Curtis P-40. It was not a solid model; it was built from templates with bulkheads and stringers and would be covered with silk. It was really quite a detailed kit. I had been relearning Morse code. I studied a little, would read a lot, think a lot about home, and then think about the project.

Carl had some pictures of those two Russian ships and the three-tailed helicopter that buzzed us. Once again, Tom said he couldn't give any pictures to me because I didn't have "high enough clearance." That was okay by me. If I could see Carl's pictures, I could just draw my own rendition. Tom Bleaker said the B Crew shouldn't be too interested in what happened to the A Crew and that we didn't want to show too much curiosity. My own feeling was that Soviet surveillance, particularly that close, should be of interest to everyone, including the State Department.

After dinner that night, I took leave to the steel beach to watch the double sunset again. The sea was so calm it reminded me of a bathtub. The trade winds were blowing softly that night, and the sea at the bow was phosphorescent. I noticed that the North Star was lower on the northern horizon compared to my view at home.

I had a good night's sleep that night. The next morning I did my usual three-mile walk then returned to our quarters at around 0600 hours. On the way I stopped to look at Carl's picture of one of the Soviet ships we had seen at the outset of our leg of the project. I took my number-two pencil and free drew the outline of the picture on his desk. Then I began to fill in the details. Carl was usually busy in the galley supervising his breakfast and dinner gang from 0500 to 0700 hours. Then, rather embarrassingly, Carl walked in. He said the wind had messed up his hair. There ought not to be much wind in the galley, I thought. Of course, he saw what I was doing. He didn't say anything about it, though I guessed that he wasn't as concerned about my curiosity as much as Tom Bleaker was.

Captain Tellaman came to me in the hospital that morning. He gave me a history that he'd been having trouble with kidney stones for years. He showed me one he'd passed the previous night that was as big as half an olive pit. He was asymptomatic that morning, though. I asked him to let me record a urine specimen and if he was having any pain now.

"No, but it felt like I was going to explode when it was coming down last night," he said.

"Do you usually have pain medication available for this in your quarters?" I inquired.

"Yes, but the last time was months ago, and I guess I'd run out and didn't renew before I left Lahaina."

He told me that if Global Marine knew of the problem, they'd not let him sail, so he challenged me to keep quiet. He just wanted me to know in case it happened in the next few days again. There was nothing much I could do for him now. I decided to keep his secret, at least for the time being.

Jack Thiel was going bug-eyed making that P-40 model. It was awfully up-close, meticulous work (play?), but he was enjoying himself and still making himself available in the sick bay. The fans removed the glue smell from our close quarters so that it wasn't so bothersome. The rest of my morning was spent studying a book on electrocardiography interpretation recommended by Dr. Borden and going over notes and doing paperwork. I ran across some distressing news that morning as well: even after our decontamination efforts, we were still getting environmental readings in the moon pool that were elevated.

After lunch, I gave Keith Milton a tour through the decompression chamber area. I demonstrated the capabilities and the fine compressors and banks that we had. I also demonstrated the facilities for oxygen decompression. He seemed quite impressed with my tour.

Dinner was a Thanksgiving turkey spread that included even the little boiled onions in white sauce and fresh cranberry sauce. This meal reminded me of home since, for the next few days, we were really all just passengers in a time warp waiting for the mainland. The *HGE*'s cruising speed was only ten knots, and so this curious-looking oil derrick just plodded along through the sea about as fast as I could jog the helipad.

Thursday, September 12, dawned much like it had the day before with orange sherbet cumulous clouds on the western horizon. The sea was a slate gray and calm, with no "popcorn" wave caps. The buildup of clouds to the east, however, suggested that we might have thunderstorms that evening. I went through my now-regular routines and made myself available if needed. I did my inspection of the noise levels in the various compartments. I worked on my Morse code and read some medical literature. My roommate, Carl, rested almost all day in his bunk. He said he felt okay, but he was bored. I really didn't think he was all that well and that he was just not telling me.

After lunch I went up to the steel beach to catch some sunrays. A big white vessel came over the horizon on the port side. It matched our speed for a while as it came closer. It was obvious to me that he was really getting an eyeful of this unorthodox floating miner. Then it got its legs and raced ahead of us to disappear over the horizon. The public address system let us know that it was a Japanese fishing vessel.

At dinner, one of the communication center personnel told me that they had a conversation with the Japanese captain. He'd asked, "A very strange looking vessel. What do you do?"

"We're the *Hughes Glomar Explorer*," he'd informed the Japanese captain. "Stand by one. We're about to go into orbit." When we didn't go into orbit, the Japanese captain made some sort of comment in Japanese and pushed his throttles forward at full speed.

Sunset came earlier that night behind the distant cumulous thunder heads that made the sky red as blood. The clocks were to be set an hour ahead at 2400 hours since we were moving into a new time zone. A map was up on the bulkhead of the mess hall showing our "noon to noon" progress. The twenty-four hour distance is about the same as I can fly my Bonanza single-engine plane in an hour!

I went to a movie that night. The popcorn was good, buttery and not too much salt. Then I headed up to our bunkhouse. Carl was reading when I came in. It must have been one of those books you just couldn't put down. He read until 0130 hours.

The next morning was Friday, September 13, which seemed apropos, as we were in the middle of our transit to be out on an unlucky day, a double jeopardy for the seaman. I had a tender point behind my second molar, upper left. I noticed a lump there too, or else I might have thought it was sympathy pains or gout or something. I decided that if it got worse, I would have to get one of the med techs to take a look at it. I balanced my concern with thinking the knee was better and thanked God for that. I was back to jogging my three miles, but alternating direction each day. No complaints from the sleepers below deck either.

Wayne Kaplan came in with an abscessing second left upper molar. Maybe mine was just sympathetic pain for Wayne. Anyway, I gave him a shot of penicillin. I wanted to suppress the infection first. If it flared up again before we got home, I would have to extract it. I've practiced the procedure on plastic models, but I had never done it for real before. Next was Bob Gleaner, who came by doing his semi-annual controlled substance audit. Our records were all in order for the narcotics and prescription medicines, but there were three fifths of medicinal brandy that we couldn't account for. I thought back but couldn't recall anyone having any use for medicinal brandy since I had come on aboard.

I looked in the cabinet where the brandy should have been stored and couldn't find any. I looked in the logbook for dispensing. It showed that three bottles were used on August 16. No wonder I never had gotten a full briefing from Dr. Borden that day! I wondered if the empty bottles were in the locker in my stateroom.

The sea was flat and calm. No swell and no chop. It looked like Lake Cachuma in the Santa Ynez Valley at home. The ship's engines slowed and then idled. We slowly came to a full stop. I learned that we were disposing of the last of the "black" material from our salvage. Later I was told that what we had done could be considered piracy. I thought to myself that there was always a few spoilsports in the crowd and hoped that we wouldn't be called up for this last deposit.

While we were stopped, I tested Friday the thirteenth to its limit. I climbed the central tower. It was not really all that dangerous a climb, but the view from the top of the main deck was spectacular. The crown of the derrick was substantially higher than the tips of the docking legs. After taking a good look around, I scratched my name into the steel plate looking down on the elevator-pipe sub spinner. All in all, it was an adventurous Friday the thirteenth.

Next morning I was up at 0600 hours and completed my routines of exercise and went for breakfast. It was a quiet mess hall that morning, maybe a dozen in all at the tables. I noticed that we had about ten late-teenagers on board as room stewards. Most of them were longhaired with beards, tall, gangly, and soft spoken. I was sitting alone at my table. One of these stewards was sitting alone at a table a couple seats removed from me. His tray had helpings of granola, milk, grapefruit juice, and an apple. His elbows were resting on the table. Unobtrusively, he clasped his hands with fingers intertwined, rested his forehead on them, and was in silent prayer, a living Norman Rockwell painting, in my thinking. My heart leaped to defend the young people who are misunderstood, and I thought of my own son, Randy, who had just turned twenty years of age and how this could have been him seated across from me. It was a real spiritual lift for me. I was not aware whether we had a definite chaplain on board. I was sure that some of the denominations had acting spiritual leaders on board, but I didn't know who they were.

After breakfast I went over to sick bay thinking I was going to perhaps extract a second molar.

Apparently, the penicillin was working well, because Wayne hadn't come back. If it subsided, it was only a temporary cure. Sooner or later it was going to recur. I resolved to advise him again of the need for follow-up care next time I saw him in the mess hall.

At 0810 hours, Carl walked in. I didn't think he was feeling too well. His history was non-specific. It wasn't claudication, or shortness of breath, or gastrointestinal distress. It was just his vague sense that he didn't feel well. I gave him a verbal placebo of encouragement and put my hand on his shoulder. Carl seemed satisfied with this, and I didn't press him further. He would tell me more specifically when he felt comfortable about it.

We were dead in the water for most of the day. The *HGE* had another engine with trouble. Well, the truth was that we had the well flooded with about five feet of water. The ship's roll sent the seawater in it crashing back and forth and sent foaming spray to the catwalk! No one told us why there was water in the well or what the eventual resolution of the problem might be.

The sunset was its usual brilliant and twinned. The movie was not that interesting to me that night, so I went to bed and read. I think I was asleep by 2000 hours.

At breakfast on Sunday, September 15, I was told that we had gotten underway the night before, about an hour after I'd fallen asleep. The well was dry, but we were told it was still not completely clean. So that was why they had flooded the well. It apparently didn't do the job.

One of the youngsters came to me in sick bay that morning. He wanted to know if I had a bottle with skull and crossbones on it. I showed him one on a bottle of chloroform lotion (the eyes, nose, and mouth are black). They wanted to copy it so they could make a Jolly Roger flag to fly beneath the Summa Corporation's flag as we entered

Long Beach harbor. "It's not a very good idea," I suggested.

"How about painting a small, red submarine on the stack?" he offered.

"No, that's not a very good idea, either," I replied. He'd seen the skull and crossbones, so I would just have to wait and see what happened when we hit Long Beach. No sense in passing any potential prankster project on to the rest of the crew.

Chapter XIV

THE LONG JOURNEY HOME

Monday, September 16, found us still in the middle of the ocean, but the mainland was pulling strongly. The freezer was cleaned out, and some of the stuff I found had touches of freezer burn on it. It had been in there since Chester, Pennsylvania. On my previous inspections there had been just too much stuff to see it all. Now that we were cleaning out the freezers, some of this stuff was finally coming into view. What a shame.

I ordered three cases of frogs' legs, thirty pounds each, to be given the deep six. There were also burned hamburger patties, pies, a dozen fifty-pound slabs of frozen bacon, seventy pounds of ham hocks, and eight hundred pounds of assorted pork chops and roasts. It all went over the side. It would have still been edible if it were in my freezer, but the law said if it was even tinged with burn, it had to go.

I was on deck with the stewards when it was all thrown overboard. It hurt to see it all tossed. Then the circus began. Almost instantly, there was a swirling in the crystal water. Like the Biblical Genesis, almost instantly there were thirty-foot great white sharks from beneath the ship attacking the sinking frozen chunks of meat. I saw one of the predators chop a frozen fifty-pound slab of bacon in two. The water was swarming with the beasts. It was a feeding frenzy!

Suddenly the crystal clear water was tinged with red. The sharks were attacking each other in the frenzy. What a horrifying display of nature's brutal power! Within ten minutes the show was over. Almost a ton of frozen meat had become hors d'oeuvres for a group of destroyers. I wondered how long they continued the cannibalism of their own.

I hibernated most of the day. Jack Thiel was coming along fine with his P-40 model. He said he intended to fly it off the steel beach into the ocean when he finished it. It seemed like an awful waste to commit it to the briny. It really was a very precise model of the old Flying Tiger's famous plane, with about a three-foot wingspan.

Sunset that night was almost too beautiful. The center of our solar system was a huge red ball sinking through a lavender sky. It was a red-orange streak across a cobalt blue ocean. Some tiny white fluorescent clouds way up high turned orange even after the horizon was dark.

I watched a movie that night. Afterward, I mused that it seemed strange that when there was no chance of getting home, one didn't even think about it, but as the time neared, like now, it just dragged. Also, it was strange how world politics or sports (the America's Cup race, for example) became unimportant if one didn't hear of their progress from day to day.

On Tuesday, September 17, it had been just five weeks since I had picked up the tickets to begin this adventure. That morning at breakfast, Bill Olmeras told me the level of contamination that remained in the moon pool wouldn't quite measure a millimeter on a side. Even so, it represented a nuisance that couldn't be ignored. I had a feeling we would have this coming around periodically to haunt us for the next year or so. It is an awfully heavy material.

I went into the hospital and inventoried the narcotics and barbiturates in the pharmacy. The numbers agreed exactly with everything on the previous inventory. Strangely, I didn't find anything that showed that Dr. Borden had inventoried the barbiturates at all. Also, I didn't find any record that there was ever any medicinal brandy in the stores. There had been a plethora of anecdotes though about the party on the night of August 16 when the crew changes took place. George, the med tech, confirmed that there were three empty brandy bottles in the trash the next morning. I thought Bob Gleaner was looking for the evidence, and he thought we were lying to him. Sorry about that, Bob.

It was cold and drizzly that night. It was amazing to me how much shorter the days had gotten as we came north. It was a nice sunset that night, but too cold to stay on deck to watch. I did run my normal distance that morning and had no problems with the knee. It was too cool on the steel beach to get any sunrays.

On Wednesday, September 18, I got up and hit the helipad. It was really quite cool that morning. We were in an early phase of shut-down mode, and there wasn't much to do. I read a lot and did a couple of hours of continuing medical education (CME) from my family practice journals. To maintain our licensure in California, we were required then to log fifty hours of CME annually. It was too dark, overcast, and cool for any sunbathing all day. It was the first time that had been true since I came aboard.

We were buzzed by a plane that day. I didn't see it, but the guys on deck said it was a twin-engine Cessna and had one engine feathered. They may have been trying to conserve fuel. The Cessna 320 had no problem maintaining near-sea-level altitude with but one engine. I was not sure how much fuel that would save, though, and wondered why the pilot was doing that.

At the supervisors' meeting that morning, we were told that there was picketing by the Union at Pier E in Long Beach. The end result of that was that we were not going in on Friday morning. We altered course to 117 degrees and almost shut down the screws. We were going in at 0100 hours Saturday so that the vans could be unloaded safely in the dark. It was a really cold, drizzly evening that night.

Oh, yes, we had a pigeon resting on the middle antennae array hitchhiking again. He was active and apparently healthy. He came aboard right after lunch. I wondered what he was doing way out where we were.

On Thursday, September 19, our crossing was almost completed. It was a great day to hibernate. The wind came from the north, which made for choppy, popcorn seas. There was no activity in the moon pool, and the pigeon was still dominating the middle antennae array. The guys had been offering him sunflower and poppy seeds.

Some of the crew remembered the last time that we had picketing around ship. The unions weren't aware that we were a US government vessel. They were picketing Summa Corporation and Global Marine because it was obvious the vessel's crew were nonunion. Some of the tricks of the pickets last time included nails in the road to produce flat tires to crew and vendors. There were even instances of slashed tires of some of the crew's cars. A few people were actually intimidated with threats of violence. This seemed to me like a hell of a way to get people to vote for your cause.

That night after dinner, we had a crew party with some vodka punch! It was terrible. It looked like coffee. I had two Dixie cups of the celebration mix. I couldn't tell afterward if the vague cerebral uneasiness was due to the color, the taste, the booze, or just plain seasickness. No, it couldn't be seasickness. I am blessed that way.

I hit the sack after the movie, a shower, and some reading. Then dripping water! There was another leak in the bathroom overhead. This was the only ship I had ever known that could sink from the top down.

On Friday, September 20, I awakened with channel fever. It seemed to have its grip on the whole crew. I didn't see anyone really smiling at the breakfast tables. It was foggy and cold with moderate seas. We were hanging just off San Nicolas Island. Being that close to the mainland, the television reception was great, the radio was a lot better than at sea, and we would soon be getting mail. Our pigeon had left us by then. I guessed that he saw the island and would rather have a worm than sunflower seeds.

There was action on the A frame. The crew was painting it, but not the moon pool. We began securing sick bay so we'd be able to go ashore after tying up. Somehow the day passed by. The meals were just as delicious as ever, but there was not the usual camaraderie among the diners. Everyone seemed unusually sober. I was in bed by 2130 hours. Oh, yes—the plumber came up and put another Band-Aid on the water pipe above the shower.

I woke up at 0030 hours on Saturday, September 25. We were underway, but I'd heard the engines slow, and that was what had awakened me. I was thinking that maybe we were coming into port. I quietly dressed and left the room. Carl didn't wake up. I headed for the steel beach first. We were in some very dense fog. I couldn't see from one end of the ship to the other. I then headed for the bridge.

Captain Tellaman welcomed me, but it was implied that I should stand out of the way and keep my mouth shut. The radarscope was just in front of me. I looked over the shoulder of the operator and could see the reflection of the breakwater. It also showed a little pilot boat coming out of the harbor to meet us.

This was truly an amazing ship. We came all the way in to Pier E under our own helm. Then the tugs turned us end for end, and we tied up with our starboard side. Our thrusters could have turned us around just as easily, but they made an awesome amount of noise, and we were really trying to keep this as silent an operation as possible. Luckily, there were no pickets there.

The first vans and boxes were offloaded onto trucks before the gangway was even down. I wondered who won the anchoring pool. It was 0215 hours when the first lines went over to our ship. How exciting! There sure was a lot of silent activity going on the pier when we arrived. Bill Hazelman was on the dock. They had him on digitalis and quinidine, I had heard. Strangely, Dr. Borden was not there. I went back to bed at 0400 hours. Carl was just up and was getting dressed. I told him about the excitement of watching the radar bring us in when we couldn't even see the bow of our own vessel! It sure felt good to be home again.

I was up again at 0615 hours. No exercise program for me that morning. At breakfast J. P. said that Dr. Borden wasn't coming in, but that I could go home. Dr. Del Hines was to be at Harvey. He was to cover any emergencies that might come up at Pier E. Bill Olmeras was going to the apartment and then to the Long Beach bus depot. I decided to ride with him. It was a $2.40 fare to LAX. Golden West Airlines had me home by 1100 hours. It had turned out to be a beautiful day after the fog burned off. I was told to have a great weekend getting reacquainted with my family and that they wanted me to come back to Harvey on Monday, September 23, for debriefing.

Chapter XV

R&R AND THEN BACK TO THE SHIP

The debriefing at Harvey on Monday, September 22, was pretty simple. It was mostly for the benefit of those who'd stayed there as supports while the rest of us were out having the time of our lives. The plan called for us to continue with our current programs for the time being. We would be notified after engineering completed the retooling for the ship and the CV and we got final approval to return to the site.

The clinic had suggested that I take a week for holiday to get reacquainted with the family. After the project Sea Spider that I had been on, Laura and I had had a two-week holiday visiting all the outer islands of Hawaii. This time we opted to take our son Randy's Datsun pickup with camper shell and just go camping.

We were off on the morning of Tuesday, September 24. Our visits included a night at Montano de Oro State Beach Park. It was a beautiful campsite. About 3:00 a.m. I awakened to feel the camper gently moving from side to side. There was noise coming from the cab. I couldn't see in the darkness, but I cautiously picked up the flashlight, slid out the rear door of the camper, and carefully picked my way forward on the driver's side of the pickup. The door was ajar. I switched on the light. Two pair of eyes flashed back at me. It was a raccoon! He'd turned the handle of the door and was into the Styrofoam cooler. He'd gotten the carton of eggs! He dropped most of the eggs and disappeared into the darkness before I could even turn around. Laura was waiting for the explanation when I climbed into the back of the camper shell. By the time I finished telling Laura the story and we were settled back down in our sleeping bags in the cozy camper shell, it was near dawn on the eastern horizon.

We made a simple breakfast of what was left of the eggs and made up bacon, pancakes, coffee, and orange juice. We continued up the California coastline through San Simeon and saw Hearst Castle.

We then headed across the state for the Sequoia National Forest. We took our sweet time driving along and just visiting with each other after all those weeks of being apart. There were songs to sing and thunderheads to watch develop over the rugged Sierra Nevada mountains. Oh, and lots of huge granite boulders and tall trees to behold. We were into our campground by 3:30 p.m. on Wednesday, September 25.

We spent a restful afternoon with more small talk and me reliving the adventure of the project with her. She had brought some nice porterhouse steaks for us in the ice chest. We barbequed those up and then sat around our campfire with a full moon as our light. It was a very restful couple of days in the cool and quiet of the forest.

When we returned home, there was a lot of mail to catch up on. I received a call from Dr. Borden from Los Angeles. He told me, "Hope you had a great time, Jack. We have everything under control down here. We'll call you when we need you."

Dr. Borden did call back on Thursday, October 10. He advised me that we would be at sea the following week. On Saturday, October 12, Hal Sampsel called. It seemed they had intended to be at Catalina Island the week before for the de-mating phase, but that would instead occur the next week. He told me he would ring me when things looked like they were ready. Meantime, I was to carry on with my usual clinic activities.

Hal Sampsel called again on Monday, October 14, at 2315 hours and woke me up from a sound sleep. He told me, "We're on hold until 1700 hours tomorrow, Jack. Come on down to the PO tomorrow morning before noon. We'll provide the agenda when you get here." They'd rather that we not discuss it over the phone, apparently.

I left home at 1000 hours on Tuesday, October 15, and drove to the PO at Summa Corporation in Los Angeles. Both Dr. Borden and Dr. Flickenger were waiting for me when I arrived. We drove on down to Pier E in Long Beach, where we interviewed Dr. Stan Baxlow, a general surgeon. Dr. Baxlow was to come aboard the *HGE* six days a month after the CV was removed. He expressed no desire to go to sea at all. He would do just fine for a fill-in medical officer for the ship while we are tied up at the pier.

We drove back to the PO, and I was briefed on the problems that we were having with the barge. It seemed there had been an electrical failure, and then a pressure hose from the control room failed. The barge just wasn't ready to go. After two holds last week, the word was that we'd be going out at 1400 hours that afternoon. Then it was extended to 1700 hours, which was when Hal Sampel had phoned me at home. We were told that we were on hold again. So we planned to go to Dr. Borden's going-away party at Hal Sampel's home instead.

It was a studio apartment within walking distance of the PO. Dr. Flickenger, Hal Sampel, George Baxter, Doug Stark, myself, and several other guests imbibed in a few toasts with adult beverages, and then headed for a local hamburger restaurant with some acclaim. It was a nice party.

Afterward, I drove out to the *HGE* at Pier E and arrived at 0000 hours. I found my way to my old bunk and wondered if that dumb pipe on the shower would break that night as I climbed in. Ron Hamler was the chief steward for the A Crew and shared the room with me. Ron informed me that breakfast would be at 0600 hours. We had word then that we'd be holding until 1400 hours tomorrow.

The room was familiar, as was the bed. I didn't even remember my head hitting the pillow. It was pleasant to be back aboard. I awakened at my usual time but rested and reminisced, since I didn't have any exercise program to perform.

Jack Thiel was holding forth in the medical department. Jack was disgusted with the mission. I thought he really wanted to go to Valdez, Alaska, with Dr. Borden instead of back out to sea. I did my reports of the Lahaina leg of the project and saw a few patients in the sick bay. The divers were to be a new crew this time around. I was delighted to find that several of them were from GM's "rent-a-frog" cadre. Several of the crew were already at Catalina Island, but Bill, Mike, and John were there with me on the *HGE*. They came by the hospital to chew the fat. It felt like a hometown again.

We were then advised that there were a lot of manganese nodules available as souvenirs at the gangway. Apparently, there were a lot of them that came up with the TO. I got a few of them to take home. I broke a couple open and found fossilized sharks' teeth in two of them. Laura had one of them made into a pendant with gold mount that I wore for many years after completion of the project, along with the two medallions that I wore throughout the whole exploit mission.

The hold went on to 1700 hours. At dinnertime we were informed that the hold was to go on indefinitely. J. P. was aboard, but only momentarily. He told me I was to go home until further notice. I gave one of the divers a ride to LAX. He was to catch a flight to go home to Oahu. I arrived home at 2100 hours. Dr. Borden was driving his car to Klamath Falls, Oregon, where his plane was tied down. He was then going to fly to Valdez.

Chapter XVI

GEARING UP FOR ANOTHER MISSION

After the excitement of the project at sea, it was a little difficult to return to the mundane, but my activities required at the clinic and the medical society offices at the national, state, and local levels kept me occupied. CHEC had come under severe criticism by the local medical society's ethics committee for "advertising" multiphasic health testing. There was still the DGMSO's contract work that kept me proficient in my diving medicine. All these things contributed nicely to the salary that the clinic awarded me.

On Wednesday, October 30, Dr. Baxlow invited me to visit the *HGE* to inject one of our divers' elbows. It was a pleasure to revisit the unique ship and visit with some old friends. Tony's elbow had taken a bump the previous week, and he'd developed an olecranon bursitis. On Wednesday, November 20, I drove down to the *HGE* again. The B Crew was back on board. On this occasion I escorted my mother, Mary Kae Rutten, and Laura through the less restricted public areas of the ship. They were impressed with the massive size of the ship and its equipment.

Dr. Baxlow had been studying diving medicine. He'd never been in a chamber before. He invited me to come down on Wednesday, December 11, to introduce him to deep diving and qualify him for his oxygen tolerance certification. My son Raul and I drove down together on that day. Raul had qualified for the marine technology program at Santa Barbara City College and was scheduled to take courses to qualify for commercial diving certification. He was very impressed with the ship too.

I took Dr. Baxlow to the equivalent depth of 112 feet of sea water in the chamber. Then we brought him up to sixty feet and had him remain in the chamber. We ran him through the standard oxygen tolerance test. This involved the subject being pressured to a depth of sixty feet of seawater equivalent pressure for a period of thirty minutes breathing pure oxygen. For this test the chamber must be specially equipped to dump off the exhaled breathing gas overboard into the ambient atmosphere.

The use of pure oxygen is always associated with the possibility of accumulating an oxygen atmosphere greater than the 20 percent normally found in the chamber. The risk of a fire disaster in such an environment is substantial. Both of our chambers were equipped to ensure all exhaled breathing gases would be removed from the chamber. Dr. Baxlow tolerated the test with no problems, and I certified him for successfully completing the test.

Until Wednesday, January 29, 1975, there was only the routine work at the clinic and trips to the SCARF range at Santa Cruz Island with the DGMSO people to keep me occupied. Then I received a call from Dr. Flickenger. He wanted me to meet with him in Los Angeles the following day to discuss the project. It seemed that Oceanus would be gearing up for the "new" project, Azorian II.

I flew down to Los Angeles with Oceanus divers Dick, Bob, Lloyd, Swede, and Davey on the Queen Air flight on Thursday, January 30. At that time we were looking for the activity at Catalina Island to begin on or about April 1, 1975. It was like a snifter of great brandy to me! Yes, I will be available, I told them. When I returned home, Laura was just as enthusiastic about a renewal of activity as I was. Of course, she knew only that it would be happening. She was not told of the time of the activity or the extent of the planned mission.

On Wednesday, January 8, 1975, the *Los Angeles Times* had an article, "CIA Reportedly Contracted with Hughes in Effort to Raise Sunken Soviet A-Sub." Staff writers William Farr and Jerry Cohen wrote the byline. It was buried on page eighteen of Part 1, which might have been a concession of good will to the agency. The gist of the story was to explain that documents stolen from the Hughes Summa Corporation offices on June 5, 1974, would confirm the association of the company with some governmental office. The article also had a *Los Angeles Times* photo by William S. Murphy of the *HGE* tied up at Long Beach Harbor.

Then on Friday, February 28, the *Santa Barbara News Press* carried an article on the Summa Corporation's *Spruce Goose*. It had an Associated Press laser photo above the article that showed the famous plane airborne in 1947, flown by Hughes himself, in the Los Angeles Harbor. The story on page A-4 stated that the plane would be cut up and distributed to eight museums and the Smithsonian Institution.

My next trip down to Pier E was on Wednesday, March 12. They were loading pipe on that day. It was nice to see the "Amazing Lady" fired up again. The entire operation gave the low profile impression that would be necessary to keep the media from becoming too intrusive.

My new med techs would be Tom Dole and Jim Barlough. Tom was in his early thirties and had been aboard for some months as caretaker of the hospital. Jim was well trained, probably in his forties, and a seasoned emergency room technician. Tom's current hobby was building a model brass cannon. He was very good at this too. I felt confident that the three of us would be able to handle any problems that might come up during our sea trials.

Dr. Flickenger telephoned me on Tuesday, April 1. The PDS now listed Monday, April 14, as the date for *HGE* Sea Trials. We were to be doing the CV transfer and mating process until April 27. I would be the medical advisor/consultant by title. Dr. Borden was on a summer voyage, and Jack Thiel was no longer assigned to this part of the project.

My good friend from both the A and B Crews, George Benko, would also be coming back on board for the sea trials. Bill Bordera and Mike Sundberg were divers with med tech experience. Dr. Kayne was a local anesthesiologist who would be giving a refresher seminar to the med techs.

Mr. Leo Bancroft would also be aboard. He was a DGMSO's director and an experienced diver. GM would be supplying us with most of our divers for the Sea Trials. A few of the Oceanus divers would also be along. Mrs. Betty Gilmore, RN, from GM, would be our liaison to keep Dr. Flickenger in close contact and to update Leo Bancroft. Mrs. Gilmore would also arrange for a cardiopulmonary resuscitation program for the divers.

On Tuesday, April 1, I drove down to the Long Beach Harbor to check in with Tom and Jim. Hal Sampel was on board too. He was being reassigned and wouldn't be with us. Our sail date was now set for Monday, April 7, but unfortunately the PDS slipped some more. The only thing germane to the project was a *Los Angeles Times* article on "Anatomy of a Secret Mission." I was glad to see that the news media hadn't forgotten us.

The PDS slipped some more. On Wednesday, April 9, I visited Harvey again, flying down on Golden West and back on United Airlines. Tom and Jim had everything in "go" at the hospital. The PDS now said the 15th would be D-day. I received a call from Dr. Flickenger on Monday, April 14, that confirmed it. He said that sailing would be later on in the afternoon and assured me that driving down the morning of the fifteenth would leave plenty of time before departure.

Laura, Randy, Raul, and I had a nice family party the night of the fourteenth. The boys, twenty and seventeen years of age, understood that I'd not be gone more than two weeks, and that I would be in the western Pacific waters the whole time. On Tuesday, April 15, I was up at 0700 hours for "chirps" and coffee. Laura fixed me a sliced tri-tip steak sandwich on sourdough toast along with half a grapefruit for breakfast, and I was almost trained for the meals I knew were coming soon on the *HGE*.

I walked around the yard and gardens and made sure all the systems for their care were in place. I left reluctantly at 0820 hours. It normally took two and one half hours to drive the 145 miles from front door to Pier E and park. There was a gradually increasing drizzle coming down rather ominously, I thought, as I made the drive to Long Beach that morning. I filled up the tank with gas and added a quart of oil for $4.80, and then drove to the parking lot adjacent to this remarkable ship.

Tom Dole was on duty. There were new badges for us at the facility office. I spent most of that afternoon putting the GM divers' physical examinations together. I also reviewed some of the crew's records. I noticed that we had one crewman with an abnormal ECG, another with hematuria, and another had bilateral inguinal hernias. Finally, I noted a pipe handler whose routine urinalysis showed acetone and protein. I conferred with the supervisors of these crew members. Were they absolutely essential to the project? Considering the qualifications of these men and the time-consuming routine for security clearance for new personnel, the supervisors waived the restrictions and accepted the medical alternatives. That's what I needed them to say.

Divers from GM whom I had dived with for years included Dan Mandani, Dick Holding, Paul Santi, Dick Dart, Larry Manley, and Dick Salmons. Oceanus divers from both A and B Crews included Bill Bordera, Tony Arami, and Mike Sundberg. We all had lunch together and talked about the project, of course. The GM rent-a-frogs were more impressed with the salad and condiment bar, it seemed, than the conversation topics, but all in all, there was a feeling of closeness and camaraderie present among us already.

After lunch I checked out the steel beach. My heart jumped a little when I saw Bill Hazelman, of apparent atrial fibrillation fame, gazing at the harbor. He reassured me that he was not going out with us. He was to be at Harvey throughout the trials, though. He hadn't been having significant trouble with his cardiac problem, he said, and his medications had it under good control.

Jack Dominic was also on board that day. He'd had a significant myocardial infarction (heart attack) only a few months before. I breathed a sigh of relief when I was told he'd been offered shore duty for the trials, but he refused it and would be going home.

One of the specialists was Randy Martinique. Randy had been asked by his supervisor to come in to see me before we sailed. His history was scary. It seemed that last August, he had noticed a scab on his right scapula that bled. On December 1, his girlfriend convinced him to have a doctor look at it. The doc, a neurosurgeon, had biopsied it. It was a melanoma! On December 8, he was in the hospital for five days for wide resection of the area and removal of axillary nodes. They were positive for tumor. On Tuesday, March 25, just a month or so of his assignment, he was back at the hospital for a follow-up surgery, and two more malignant nodes were found. Apparently Randy was involved in the communications center and had specific technical qualifications that were hard to replace. I waived his history as a disqualification but resolved to talk with Dr. Flickenger about his case. Since we would not be going to an area out of chopper range, I didn't think we would be in real trouble in keeping him on board. Maybe we would have to use the steel beach for the purpose it was designed for, if worse came to worst. I asked my Friend up above to keep His hand on Randy's shoulder.

Chapter XVII

STRETCHING OUT THE LEGS IN ROUGH SEAS

I had a rather fitful sleep during the night. I felt well rested in the morning but woke up a lot during the night. We left Pier E about 0000 hours on Wednesday, April 16, 1975, and sailed into the outer harbor. At 0600 hours, my new roomie, Ron Hamler, left for his assigned duties. I didn't hear him go, so I must have been sleeping soundly when he got up.

Ron was the chief steward from the A Crew. He was on duty for the recovery phase of the project too. He was quite a bit younger than Carl Atkinson but just as efficient and dedicated. His lifestyle was a duplicate of Carl's. I knew there wouldn't be a problem adjusting to our routines.

I was up to the steel beach for my exercise that morning. It was a beautiful day for Long Beach. Big cumulous clouds had built up on the mountains from a tropical disturbance off Baja California, Mexico, which caused the smog to be blown away. It had rained during the night, so the air was amazingly clear for this area. I resolved to enjoy the view because the smog would soon return and obscure the nearby mountains from sight. I did my run/walk around the perimeter of the helipad alternating directions at the midway point. I felt no pain in my knees and was very glad for that.

There was big news that morning at breakfast of the death of four movie stars that past week: John Conti, Marjorie Main, Fredrick March, and Larry Parks. Also, Phnom Penh, Cambodia, had fallen, and the Vietnam War seemed to be winding down. I was determined to compliment Ron the first time I visited with him.

On the way to the hospital, I stopped on deck and watched the news media helicopters flying around us. I imagined they were not going to allow the usual headlines, like "Glomar Explorer Vanishes," to show up in tomorrow's papers. All was quiet that morning in the hospital, so I decided it was a good time to do the narcotic inventory.

At 1200 hours, the engines came to life, and we began to move out of the outer harbor. We crossed the channel until we were about halfway to Catalina, and then we took a westward direction toward Santa Cruz Island. All of the Channel Islands came into view as we moved toward the big island of Santa Cruz. It was so clear that we could see trees on Catalina Island from thirteen miles away. The sea was fairly smooth, with a low swell and almost no chop. The sun made it feel warm as soon as it came from behind the occasional cloud, but the chill factor was high in the wind. My extra jacket felt great!

At dinnertime, we were about thirty miles south of Anacapa Island. Ron's menu included New York steaks and Alaska king crab, and I enjoyed both. It was like being on one's own private yacht, with magnificent cuisine being served when you were ready.

The television monitors in the mess hall showed us videotape from the evening news. There was a great shot of the *HGE* leaving the harbor. The news anchor was speculating about our activities and ultimate destination. Some "eye witness" woman on the local affiliate for NBC (KNBC) gave a description of the "claw" of the CV, which she claimed she saw while we were in Lahaina. I am sure she was lying, but what am I to do.

The pipe well was about two thirds full. We were supposed to rendezvous with the DGMSO's vessel, *Swan*, that night. The *Swan* had been my home on many nights at the SCARF range south of Santa Cruz Island. She was also my home when we were at Catalina Island to witness Hannes Keller's first-ever one-thousand-foot ocean dive. I was looking forward to seeing her again. During the night, we tested the engines, forward to reverse with a lot of shuddering, creaking, and groaning noises going on and things falling off of shelves. This was enough to make for a restless night.

On the morning of Thursday, April 17, it was quite cold on the helipad. The wind was blowing at thirty-five miles per hour, and there were huge seas. We went no closer to Santa Cruz Island than about five miles. Then we went back out into the basin for "station keeping" (maintaining our position without anchors). It was amazing how the ship was able to do that in spite of the giant wind and seas.

It was med tech Doug Stark's birthday that day. Poor Tom Dole was seasick again and came in looking quite miserable. My work entailed physical examinations on those members of the crew who'd come aboard without a recent exam on file. One young man was highly resentful at my inquiry as an invasion of his privacy. It turned out he was a three-year naval officer on whom the crew had already hung the appellation "Lt. Fuzz," since his personality was so like the character from the comic strip *Beetle Bailey*.

On Friday, April 18, I noted that it was the two hundredth anniversary of Paul Revere's famous ride. It was less windy, and the seas were subsiding. My lingering cold acquired just a day or two before was better, but I had a tickling cough, productive of greenish sputum.

Merlyn Carbotti came to sickbay that morning. He had been in yesterday for "back pain." He had denied dysuria then, so I gave him Robaxisal. Today, he was admitting to two episodes of gross hematuria in the last month or so. The pain was now on his left iliac crest. His urine showed only oxalate crystals. There were no red blood cells in his urine. Tom took a kidney, ureter, and bladder x-ray at seventy-two inch, twenty milliamps, and eighty kilovolts peak exposure. It showed nothing, so I repeated the test. It was a good picture, but I didn't see any kidney stones. Some stones are radiolucent, though, and don't show on an x-ray. I noted that in his chart, and I let him know that we needed to watch for further complaints of pain or discomfort.

After lunch was mail call. The chopper came in despite the still present wind condition. The pilot apologized for not making it in the day before. It really was tough to make a landing on the steel beach. I was pleased with the mail call and said, "Thanks for the letter, Honey." It was station keeping all day that day.

Just to confirm a suspicion of the news media, we did indeed have a few hardcore porn movies in the library. They had one playing that afternoon. I stopped in to see about a half a reel of one of them. The principles didn't have much talent for acting.

Scuttlebutt was that we'd be in Catalina tomorrow. I only did about two and a half miles on the steel beach that day. My legs weren't up to the unbalance of a rolling deck.

On Saturday, April 18, we were still station keeping when I hit the helipad for my exercise program. I held it down to three miles, although the sea's roll was minimal. By 1100 hours, when we moved off station and headed for Catalina, there was almost no swell and no chop. There was brilliant sunshine with only a very little wind, but damn, it was cold!

At lunch I talked with Tony Arami over corned beef and cabbage. Tony was a short, stocky Italian with a cocky disposition and a voice that sounded like Mario Lanza for about five bars of opera. Tony was the kind of character every story needs—a little man with a sailor's vocabulary. I thought the only adjective he knew was the "ing" suffix of the F-word. But he was colorful! And when he was serious, he was all articulate business. He had tattoos on his arms, and to hear him tell, he had a girl in every port. He wasn't married when he was on board with us. Tony was an Oceanus diver. He told me the divers' main complaint was that the company had been a little frugal lately, and that was why so many of the divers had left the project.

The ship stopped about five miles off Avalon on Catalina Island and began filling the well. At 1900 hours it reached thirty-six feet, and the hull doors were opened. Larry Nanley and Bill Bordera made the first dives in the moon pool. It was uneventful, and all was clear and orderly. Now we would be laying pipe for the next few days.

That afternoon we had an inspection by Captain Grafton. He was eagle eyed and thorough, but not a white-gloved zealot. We toured the power generators. There were five steam turbines that drove the ship and supplied power for all the electrical requirements of the ship. Captain Grafton told me that the capability of these generators could light the entire city of Long Beach. Usually, only two were running at a time, unless we were laying pipe.

The fuel cells for the ship were in the double hull. She held enough diesel fuel to run the ship half way around the world. The captain asked me if I would escort him through the hospital and sick bay. I was delighted. I gave him a thorough tour of the facilities. The med techs, as usual, had everything clean and in order. I was very proud of them. He thanked Doug Stark and George Benko for their hospitality. We got a 4.0 grade.

After dinner I went up to the steel beach to watch the sunset, and then back to Menninger's book to lull me to sleep. I think it was about 2130 hours when I turned out my reading light.

There were lots of sail and powerboats that came out to ogle us on the morning of Sunday, April 20. It was a calm, sunny, warm Sunday morning. We were laying pipe, but two transponders weren't working again. That made for four failures so far on this trip. At $5,000 each, you'd think there'd be some kind of guarantee with them. I am not at liberty to discuss the rate that we could lay the pipe;

however, it was at a quite amazing pace to go to the depth that we needed to go.

One of the stewards had an ingrown toenail that morning that was giving him a lot of trouble. I relieved him of the malady with no pain after the anesthesia, of course. Another crewmember, Jamie, had a recurrent sebaceous cyst on his right arm that Dr. Borden had removed last summer. I decided to store the cyst in a specimen jar of 10 percent formaldehyde for a year to check back with.

The water in the moon pool was now crystal clear. I inspected the decompression chambers and fired up the compressors. I didn't think we would need them for this exercise, but it was comforting for me to know that everything was in working order. After lunch I lay on the steel beach. It was a restful Sunday for me.

Dinner had a Mexican theme. It was carne asada with beans and rice, and all could be wrapped in a giant flour tortilla. There were also beef and cheese enchiladas, chiles rellenos, or any combination, if one was really hungry. I thought to myself that Laura would have loved this meal. She had enjoyed hot Mexican food since we had begun making flights to Baja California back in the 1960s with our own Bonanza airplane. I checked out a book from the library for reading that night. We had a good library on board, with a wealth of good reading available to all the crew.

On Monday, April 21, 1975, it was chilly, and a dreary marine layer hung over us right down to the sea. It was so thick it was almost a drizzle coming down. No matter, I did my usual run/walk. The radio news was pretty empty. Nothing was said about the *HGE*. The news media couldn't see us that day anyway, with all the fog and low clouds. I did hear that Nguyen Van Thieu had resigned as president of South Vietnam.

The crew was cycling pipe, and it all seemed to be going smoothly. Our transponders were keeping us exactly "on station." A motor whaleboat came out from Avalon and left us with some nice, big sacks of mail. How about that! I got four letters from my Honey. I give silent thanks to Laura. Her letters meant so very much to me.

There was nothing scheduled for that day at the hospital. I studied a textbook in the medical library on electrolyte balance. It was a readable presentation and made a good review of volume replacement in the event of severe blood loss when transfusions are not available.

At 1200 hours the weather was exactly like it had been at 0600 hours. The temperature had gotten up to only sixty-one degrees. After lunch I went to our quarters, took off my shoes, and climbed into the bunk to read *The Exorcist*. I only managed about four pages of the text an hour, since I dozed off for a good nap several times. It continued to be a nothing day. Dinner was the usual nice meal, with good conversation with good friends. No use going to the steel beach for the double sunset that evening. The marine layer was four thousand feet thick, and it was cold on deck. Perhaps I should have been happy that it was such a quiet day.

Tuesday, April 22, dawned just like Monday. When I hit the helipad for my exercise at 0530 hours, the marine layer was actually misting on a wet deck. It felt good running to get warm. I did five miles and still had no problems with my knees.

I felt terribly sorry for Tom Dole that morning. He'd been on a weight-watching diet and had been real good about it, too. He hadn't gotten any mail from his wife of eighteen months since I had come aboard. She lived in Washington state. When the mail boat came out from Avalon, he was really depressed when there wasn't anything for him. At that night's dinner, he had two pieces of prime rib followed by a root beer float and generally ate up a storm. Poor Tom, his weight-watching diet was down the tubes again.

The pipe handlers put down more pipe that day. They had now gotten three thousand feet down.

Everything was going smoothly, I was told. There were porpoises playing around the ship that evening. They really seemed to be having a lot of fun. A whale even went by that afternoon. We couldn't really see him, but we could see the spume when he'd blow.

I skipped the movie again that night and went to bed early to read. Ron woke up at 0530 hours to begin his duties. He didn't make much noise, but it was enough to wake me. I lay quietly staring at the ceiling. I really didn't want to engage him in conversation when I was with my own thoughts that morning.

After he had left, I got up, made my bed (although the interns were supposed to do those things), and looked at my calendar for a clue as to the time we'd already been at sea for this tour. It was the eighth day, Wednesday, April 23, 1975. Twenty months since I first went to sea with the *HGE*.

That morning they were cycling different sizes of pipe, and they were also going to cycle the docking legs. Now that was really something to watch. I picked my way up to the rigging platform high above the deck. It felt strange to stand on the gimbaled tower and watch the ship move around under you. The pipe crew really had it down to a ballet. The color of the pipe told them the changes they must make in the computer to compensate for the threading of different sized sections.

Then there was the unusual visual experience of watching the docking legs sink into the ship, like they were melting Popsicles. Gee, I sure wished I had my camera with me!

I headed up to the steel beach for my five miles and then to breakfast. I limited myself to a bowl of Kellogg's Special K and a piece of toast that morning. I gloated in my self-righteousness, but felt I just had to get my potbelly back to size.

I also enjoyed an hour of sun on the steel beach that afternoon. I reread the mail from home and noted that Monday was Laura's big day at the

Holiday Inn for the Tri-Counties' Auxiliary. I mused that being off Catalina Island sure felt a lot closer to home than Lahaina.

After lunch, I gave a lecture on CPR to the electricians on board. It was well received. I hoped we never had to use it except in practice. I had a "Resusci-Ann" model with us, but they did not have time to be certified.

The med techs told me that there was a guy named Fry, or something like that, who had declined to appear for his physical examination. We did have our problems. I didn't know if we even had a person named Fry on board. We also received a note from the PO suggesting we should be able to do seventeen physical examinations a day if we had a day's notice.

Poor Tom, he didn't receive any mail again that day. He was sure occupied with the lack of communication with his family. He was not really holding up to his responsibilities as a med tech either. I was a little worried about him.

Thursday, April 24, was a ditto of the day before. It was cloudy and cool most of the day. From the work point of view, it was just like yesterday too. The crew practiced on cycling the pipe and docking legs and timing the efforts to make them more efficient. It looked like they were really thinking that we were going to have another Jennifer.

I was asked that day by the PO to fill out a paper for the company to dispose of insurance money if I died on the job. I thought, *Thanks a lot!* They also wanted four passport photos. Well, I was not sure what they had in mind, but scuttlebutt said we'd be home in Long Beach about 1200 hours on Sunday. The high tide was at 1300 hours, so if we didn't make it then, it would be 0000 hours for the next high tide.

At lunch Dave Dickenson came to me while I was eating and said he wanted to talk with me privately that afternoon. I told him to come in to sickbay after lunch, and we'd have it to ourselves. He never showed up. Instead I had Jamie for removal of another sebaceous cyst near the angle of his eye and nose. He did show up, and I did a nice job for him.

I sat in the hospital reading the new surgery texts we had received for a couple hours while I waited for Dave. I wondered what he wanted. Then I used up some time watching them cycle the docking legs. Raising the aft legs required the starboard drive wheels to rotate clockwise and the port side's drive wheels rotate counterclockwise. It took three minutes forty-five seconds for a full revolution that raised the docking legs about ten feet.

There must have been another tropical storm down Baja California way. We were beginning to run a moderate sea. After dinner I watched the movie in the theatre. I felt that Dave might find me if I were available, but he didn't show up then either.

The wind came up on the morning of Friday, April 25, and the seas were running eight to ten feet. I woke Ron Hamler that morning because I knew we both had overslept. My Timex watch had numbers and hands that couldn't be seen in the dark. It was a good thing I did wake him; otherwise, I wouldn't have been able to have an egg and bacon for breakfast. Poor Tom showed up that morning for breakfast but couldn't even handle a cup of coffee due to seasickness. He had been given a Phenegran tablet and Dexedrine, but that hadn't helped him. He looked about as joyful as a basset hound. He also hadn't gotten any mail yesterday either.

At lunchtime Tom didn't show up again. I did have a nice conversation with our news media spokesman, Paul, from Dallas. He said he was on board only for the day. He was a great conversationalist, and we got along very well.

Dave Dickenson came in that afternoon. He really needed someone to listen to his grief. I was very glad I was available for him. His son in Phoenix was in a coma and had been for a year. He had been accidentally shot in the head. Dave had hospital bills over $75,000 and was being sued by the University of Arizona. The Internal Revenue Service was also after him, and his wife had divorced him within the last year. In addition, he had remarried last August and was struggling with marital issues.

Tom finally got a letter that afternoon and had miraculously gotten over his seasickness. We had dinner together that evening. Afterward I went up to my bunk to write Laura back home.

The seas were too big that night to close the well gates. Word was that they would wait to see if the seas subsided before they tried to close them. All the pipe was up and stowed, so there were no immediate problems with that. We were told that if the big swell continued to tomorrow, we'd move slowly over toward the shelter of Catalina Island out of the wind and then close them. We were scheduled to be back in Long Beach on high tide on Sunday. The forecast was that the ship would be involved in a mating procedure and the IST in May and June, but no one thought that there would be a summer mining mission.

At breakfast on Saturday, April 26, Tom told me that we had eighteen-foot seas overnight. Everything was going smoothly so far that morning. We had moved over to the lee of Catalina Island during the night. I was told that it was dangerous to close the gates if the seas are greater than five feet. After the gates closed, the well was pumped dry. We then steamed out sixty miles to sea to dump our garbage. Then we planned to head for Long Beach and home.

That day we put our med tech, Tom Dole, and Dick Dart, a diver, through their oxygen tolerance test in the decompression chamber. Initial descent was to fifty pounds per square inch gauge. Tom complained of ear distress at fifteen feet of seawater equivalent. This didn't improve with ascent to ten feet. His ear distress continued throughout the dive but did not seem related particularly to pressure changes. Standard oxygen tolerance test was accomplished without incident.

An all-crew briefing was held that afternoon. We were to be mating from May 5-8. IST would be May 30 to June 14 (only fifteen days), and a mission was scheduled to leave July fourth to mine "manganese nodules." The *HGE* rolled around like the GM research vessel *Swan* last night. The winds gusted to forty-five miles per hour, I was told, and the seas were twenty-five feet. We proceeded southwestward between Catalina and Santa Barbara Islands. We would proceed to about sixty miles offshore to empty our sewage holding tanks and deep-six our garbage. I had also been told that the holding tanks were processed and chlorinated to what's called "secondary treatment," about the same as the outfall sewage from the plants back home in Goleta. The garbage was containerized, weighted, and mostly biodegradable. The well was pumped dry and was clean. She really was a unique ship. There was not another like her in the whole world.

Ron was up at 0530 hours on Sunday, April 27. I lolled in bed until 0600 hours savoring this last day at sea. Yesterday's briefing, though, had assured us that there'd be more opportunities to ride the *HGE*. The seas had abated quite a lot, but the wind was still cold and piercing. I offered some of the GM divers a ride home, but when they found out I was driving a little Datsun pickup, they said they'd rather rent a car. I guessed they were not as acquainted with the Los Angeles freeways as I was.

When we reached Pier E, I decided to leave my clothes in my locker, since I would be returning in a week. Doug Stark, my med tech, gave me a big lock for my locker. Adjacent to our mooring was the cavernous hanger for the *Spruce Goose*, Howard

Hughes' huge, wooden, six-engined flying boat. The place was secured, and no one was allowed to see the world's greatest flying boat. There were twenty-four hour security guards on duty. It must have cost a lot to store that airplane. But then, the *HGE* wasn't a cheap toy either.

My son Randy's little pickup started right up and took me home at freeway traffic speed. There was a warm welcome for me at home. The family was getting used to a peripatetic husband and father. The yards and gardens looked just great!

In the Sunday, April 27, *Parade* magazine inside the *Santa Barbara News Press*, there was a great story on page 8 by Alexander Cockburn and James Ridgeway titled *The Race for Riches on the Ocean Floor*. The story was all about deep ocean mining projects and the international race with the law for exploiting the oceans.

Chapter XVIII

TROUBLE WITH THE MATING EXERCISE

Dr. Flickenger rang me at 0630 hours on the morning of Thursday, May 1, to confirm the briefing on Saturday. I confirmed that I would be ready for mating on Monday, May 5. He asked the usual questions about the quality of the staff in the hospital and the availability of medications and materials. He also inquired as to the morale of the crew. I told him that everyone seemed to be enthusiastic about the second program. I thought he had been very worried about that, but the attitude had turned around from the "no way" when it was first mentioned to "let's get on with it and finish the job" now.

It was a beautiful day in Goleta that morning. I did my mail at the office almost the whole day. CHEC was running smoothly, and my technicians were pleased. There were still a lot of the physicians who just didn't take to this new-fangled side of medicine, but all the projections by the experts said this was the coming thing. Well, we would be on the cutting edge of technology if the experts were right.

The big news that day was that the Vietnam conflict was finally over. Saigon was now called Ho Chi Minh City. I thought to myself what a waste of resources, both physical and human. The wounds to everyone would take a long time to heal.

I learned on Wednesday, April 30, that the *HMB-1* had left San Francisco and was heading for Catalina Island. On Wednesday, May 2, Dr. Flickenger phoned to tell me that Mrs. John Mackel was suing Summa Corporation and Global Marine. It had been a short three days at home, just time enough to process my correspondence and touch base with the family.

My report on the mating operations to Summa Corporation began, "Mating operations were conducted in the eastern harbor of the Isthmus of Catalina Island between the *HGE* and *HMB-1* on May 5, 1975." I left home at 0745 hours in Randy's pickup again. I dropped off my pager at the clinic and was on the road by 0815 hours.

The B Crew was back aboard for this exercise. Tom Dole was aboard and was again feeling low. Poor Tom, I felt sorry for him. Diver Mike Sundberg was in charge of the oxygen tolerance test program and was dissatisfied with Tom's chamber ride as I reported it. He wanted him to do it again. Tom, with vehemence, said "No!" I had hoped that would be the sum total of the negative part of this project but had a few apprehensions.

One of the highlights of this tour, though, was my introduction by Dr. Flickenger to a wonderful humanitarian and colleague, Dr. Beanie Coulters. Beanie had a long history of working with a certain laboratory in the Bay Area in nuclear medicine and had written many texts on the subject. He'd be along for the "Mating Game" as a guest of Dr. Flickenger. Beanie was about Dr. Flickenger's age, short and stocky, with a marvelous sense of humor. I didn't have much time to visit with him that day because of the pressing business of seeing patients in distress.

Captain Grafton came to see me in sickbay with a complaint of biceps bursitis. I gave him some non-steroidal anti-inflammatory meds to begin with. If that didn't work, I would inject it with a little xylocaine.

John Cole, an aerospace contractor, followed Captain Grafton. John had squashed the distal tip of his fourth finger on his right hand between two oxygen cylinders and had a painful bleeding under the nail. Using the red-hot paper clip treatment, I opened a hole in the nail painlessly, and the blood gushed forth, relieving the pressure. He was grateful and left satisfied.

Lunchtime provided a respite for social conversation with divers, technicians, engineers, and good friends. It was nice to have Carl back as a roomy. I guessed the two chief stewards would be changing off as new phases of the project put us to sea. That way they keep two well-qualified stewards on the payroll.

It was sunny and cool in Long Beach on that Monday, May 5. Amazingly, there was no smog in the Los Angeles basin that day. At 1830 hours we cast off and headed for Catalina Island. Carl's special for that night was veal parmesan. It was very tasty.

On Tuesday, May 6, we were easing toward the isthmus of Catalina Island. We had checked out our transponders and the ASK systems when we were on trials two weeks earlier. Everything had gone smoothly. When we were only a mile offshore, I recalled being in this same area where there was a summer camp that we had gone to with our kids years before, when I served as the volunteer physician for some 130 youthful campers. The sea-going tug *Pacific Gemini* was alongside us, as was another tug, *Kingfisher*. They were the same two tugs from San Francisco that had pulled the *HMB-1* down from Redwood City. There was no sign of the *HMB-1*, though. She was down between four buoys straight offshore and about a half a mile from the isthmus dock. The visual markers were a fuel dump on the island and a small resort in a bay, the second one north of the isthmus.

The barge *Ore Quest* was her tender. Giant compressors kept the barge supplied with air. And then there was the crane/winch barge, *The Happy Hooker*, of Big George Marvin and "Woody" Treen from Santa Barbara, with a winch that worked poorly and a crew that put the Three Stooges to shame. They backed us over the barge, took our big anchors, and made a four-point anchor to moor us with lines about a third of a mile long. Then, to get us into exact position, they snugged up one anchor chain and slacked up on another.

The well was again flooded by 1700 hours. Divers were in the pool, and the main gates were opened. It was a clear, sunny day with a cool breeze blowing. Everything seemed to be going to plan. I made a trip by small boat into the isthmus with Chuck to call Stan regarding Bob Patton. Then we went back to the ship with a guy who used to play ball with the New York Yankees.

Tom Dole was very uptight and worrying about getting to do a dive that day into the moon pool. Jim Lane of Lockheed was an old diver. He had already suited up to dive in the moon pool. As senior diving medical officer, however, I had no authorization to allow him to dive. Jim objected strenuously, but it really was my call, and since I had no documentation that said he had higher permission, he didn't dive that day. Tom got the assignment instead and was delighted.

I was supposed to be called that evening if they needed extra phone help for the divers. They were to be working that night to secure the bridle connecting the pipe stem with the strong back. They would be wearing Aquadyne helmets with phone communication with the control center. No dives were scheduled that night for deeper than fifty feet.

On Wednesday, May 7, the morning dawned sunny, windy, and cold. I was up at 0530 hours. The docking legs started down at about 0700 hours. Everything went as smooth as clockwork. I occupied myself that day mostly with studying my continuing medical education, taking off a few warts and skin tags, and cosmetic things like that.

On my rounds at 1600 hours, I stopped in the well. Mike Sundberg was showing me the central diving station when there was a "boom!" forward in the well. Foaming bubbles rose eight feet high in the well! The whole ship quivered like a frightened stallion. Then "boom!" again. There were more foam and bubbles, and the ship pitched and trembled. Then the activity slowly subsided. As the foam cleared, one could see the CV about ten feet below the surface. She was listing about six degrees to starboard. After everything had quieted down, the divers checked everything out. Luckily, no one had been hurt, but there was too much buoyancy, and the CV had floated to the top of the keyhole.

Apparently, there had been no damage to either the CV or to the ship from the massive influx of water. Cables were made fast to the starboard side of the CV. They tried to straighten her up, adjusted the buoyancy, and raised it on the docking legs until it was out of the water and secured. We all fervently hoped that would secure the potential loss of the entire project. Captain Grafton said we'd be ready to leave by 0400 hours. Three million pounds of machinery would surely smash the *HMB-1* if the CV fell.

Dave Wright caught me at dinnertime. He asked if he could see me after dinner in sickbay. I said, "Sure, let's say at about 1900 hours." I was glad I didn't put him off until the following day. When he told me of the pain he was having with a thrombosed hemorrhoid, I really felt sorry for him. When I saw the size of the thing, I was astonished! I was sure that the thing was really agonizing. It had prolapsed through the anus and was about the size of a Vienna sausage. I prepped him with Phisohex, injected it with some xylocaine, and after his acute pain subsided, incised the lesion with a #11 bard-parker blade. The clot extracted easily. There was minimal bleeding. I applied a Vaseline pad to the anus and digitally replaced the empty vein into the rectum. Dave's gratitude was genuine. I did advise him to see his own physician when he got back to Long Beach to have a sigmoidoscopic examination of the lower bowel.

Later that day I found myself a little upset with Tom again. He hadn't reordered our supply of Chlortrimeton for allergic reactions. I didn't have any for one our patients that morning. I told him we must reorder our supplies as we use them. He never wrote it down but did say he thought Benedryl was superior anyway.

I awakened at 0100 hours with the "boom!" resounding in my ears like the ones the day before only it seemed much louder. The ship lurched and twisted. "Boom! boom!" I thought, *Oh my God! Now what?* It felt like an earthquake. *If it doesn't stop pretty soon, something is going to come apart!* I thought. Fortunately, it did stop. I lay there for a while listening to the background noises of the ship. Nothing was unusual in the sounds. No alarm or bells or public address announcements. Had the sound begun again, I'd have been up and seeking an explanation. As it was, I woke up again at 0500 hours. Carl had already left, but I couldn't remember if he'd been in the bunk below when the booming sounds were going on. I dressed and headed for the deck of the well.

I was told later that the boom sounds were due to the CV falling from inside the well. The gates had been unable to close, as the CV was hanging beneath them. First we had to move gingerly off the site directly over the *HMB-1* so that, should the machine fall, it wouldn't crush the barge. The *Happy Hooker* came alongside and picked up our starboard aft anchor, but let the huge chain catenary drag across *HMB-1*'s umbilical. The crew corrected this potential disaster before it did its damage. The *HGE* then moved to a point where we could maneuver better. In another miraculous engineering recovery, the crew was able to raise the very heavy CV from within the well and close the gates again.

This scenario began shortly after 0130 hours on Thursday, May 8, and concluded at 1930 hours. The crews, and particularly the divers, were exhausted. The deepest dive was to 114 feet. Only the deepest dive required four minutes decompression in the water at ten feet.

However, some of the divers spent considerable time in the water at shallow depths because of the rigging difficulties with the CV on the docking legs. One diver actually recorded two hours and five minutes diving time. None of the divers required decompression in the chamber. All were easily handled at ten feet in the water. Thank God! They began pumping out the well. Everyone was anxious to see what the damage assessment would be when the well was dry.

During the day I had an opportunity to have a "Dutch Uncle" talk with Tom, emphasizing teamwork. After our discussion I thought he would come along all right. His attitude seemed immediately better. I thought the tension of the day's activities convinced him we really were a team. He had mentioned something about the relationship between his dad and his grandpa causing him tension. I resolved to develop more on this subject with him over time.

At dinner Jim Lane, the big, clumsy diver from one of the other suppliers, came to say good-bye. I also had the opportunity to say good-bye to another diver from the same provider who was on the *Ore Quest* when the shore boat stopped off on the way to the beach. I went in, at Ray Benkoe's request, at sundown to phone Stan and find out how Bob Patton was doing. His injury was a fifth cervical vertebral subluxation. He had quadriparesis. What a shame. His prognosis was guarded.

It had been a long, tension-filled day. We were finally underway back to Long Beach. We were scheduled to dock at 0930 hours the following day.

We arrived in Long Beach on Friday, May 9, just five minutes before we were due. There were lots of happy people to greet us. I drove Lloyd Simmeron over to the Avis Rental Car to get him a car, and then drove myself home by way of North Hollywood and my sister's home there. I had been told that my mother and other sister would be there, and they could ride home for a visit with me. When I arrived, I found that they had already left. I visited with my sister, Janice Olsen, for about forty-five minutes and then anxiously drove home.

Chapter XIX

RETURNING TO PORT

The difficult thing about this exciting adventure of being at sea was the mountain of mail that accumulated while I was off having a good time. I was home shortly after noon on Friday, May 9, and after greetings with the family and discussing the problems that had developed in my absence, I headed for the clinic and the second mountain of mail. I brought it all home with me.

The *Parade* magazine in the *Santa Barbara News-Press* on Sunday, May 11, again had an excellent article on *"The Inside Story of Project Jennifer"* written by Lloyd Shearer. There were also some good pictures of the *HGE* and the *HMB-1*. The article mentioned the front-page story published in the *Los Angeles Times* of February 1975 involving a robbery of Howard Hughes' Hollywood office on Romaine Street. It certainly seemed as though the plot was thickening!

The next twenty days returned me to the old routines. There were opportunities to work in my yard and garden, go scuba diving with Dr. Vernon Friedell and my son Raul, supervise CHEC, attend meetings of the Santa Barbara County Medical Society (SBCMS) and the California Medical Association (CMA), and participate in the SBCMS's crisis committee formation.

On Thursday, May 15, I was called out for a diving accident related to the offshore diving industry. A young man by the name of Jim Clancy had hurried his "in water" stoop at ten feet and suffered the consequences of exceeding the dive tables for decompression. He had an aching pain in his left shoulder and felt like he'd been beaten all over with a rubber hose.

We took him to sixty feet according to the Table 1-A decompression. His shoulder pain and malaise disappeared completely as we neared sixty feet. Following the table, we proceeded to 100 feet equivalent and began the long decompression at 1930 hours. At 2200 hours Jim was feeling fine, but tired. I felt he was stable, and he agreed. Our tender was able to handle the rest of the routine decompression. He brought me out separately in the outer chamber. All went well, and I emerged at 0100 hours.

I was a little tired myself when I got up at 0700 on Friday, May 16. I dragged in my at-home routines and gradually attacked the mountain of correspondence that had accumulated while I was at sea. When I did get into the office that morning, I learned that Jim had had no more problems.

In addition to my duties at the clinic, I had to get back into my active role as a local delegate to the California Medical Association. The big issue pending was to prepare for a slow-down of medical services to the public to begin on Tuesday, May 27 in protest by the CMA over skyrocketing malpractice premiums.

On Wednesday, May 21, Dr. Flickenger telephoned to tell me that we would be going back to Catalina Island on Thursday, May 29. He confirmed this and added a few more details by phone on May 27. We were scheduled to leave for Catalina Island at 0015 hours on Friday, May 30.

On Thursday, May 29, I left the house at 0800 hours and dropped off the pager at the clinic and headed down to the PO, arriving at 1015 hours. Dr. Flickenger greeted me enthusiastically when I arrived. With him were Doug Stark and Dr. Beanie Coulters. Our new fifty-year-old med tech, Jim Brandies, was also there. Jim had just arrived from Saigon with no luggage. He'd taken one of the last helicopters from the roof of the embassy in Saigon just before it became Ho Chi Minh City. Dr. Flickenger briefed us with about 150 words and fifty matches as he tried to keep his pipe going. It took him about forty-five minutes, although he really didn't say that much.

Dr. Flickenger did tell us that he'd interviewed a young surgeon in Detroit for the "mining" leg of the project. He was apparently well qualified and eager for the leg, but Dr. Flickenger said he had lots of curly hair, a beard, and a potbelly. He said he looked like movie critic Gene Shalit. After I had let him ramble on for a while, it seemed like he would ask him to come aboard from July 3 to August 15.

There was to be a crew change somewhere near the Midway Islands. Lahaina didn't want us again, and we didn't want someone else to create an incident there. We may go mining elsewhere if there was to be "hot line" interference (someone listening). Or we may see how much of a confrontation we can make with what they already had. Anyway, after the mining leg (when Laura and I should be in the Mediterranean), there would be three more legs. Possibly I would be handling all of the legs, or perhaps none at all.

After lunch of a big hamburger and a bottle of beer at the Jolly Roger courtesy of Dr. Flickenger, we continued the discussion at the PO until 1700 hours. Then, at Hal Sampel's invitation, I followed him from the Vista Del Mar Motel to the breakwater at Marina Del Rey. He parked the massive Winnebago there on a vacant lot. He was still supporting his wife and kids and so had to live quite frugally. I had a drink with him, and we took a nice walk in the foggy haze as the sun set. We shared a delicious meal, and I enjoyed the conversation immensely.

About 2045 hours I made my way back to the ship. I discovered there was a twenty-four hour hold on for setting sail. I later learned that the computer had been programmed for the wrong "Zulu" time in the southern hemisphere, the pipe slide cart wasn't working properly, and the steering motor had seized up. What can go wrong will go wrong eventually. Better now than if we were at sea.

It had been a long day. I did telephone Laura from the PO, and we had a nice conversation. It was time to go to bed. My roomie, Ron Hamler, had the room locked when I arrived, but a gentle knocking gained me entry. I showered, hit the sack, read for a little while, and then accepted the gift of sleep.

On Friday, May 30, I was up at 0600 hours. I wrote a letter to my honey and then telephoned her to tell her about the delay. It was a warm conversation, and I did my best at trying to clue her in to our plans. We were told we would be in the Santa Cruz Basin again.

After breakfast I went down to sickbay. I injected a sore shoulder for one crewman, removed some stitches from another, and then caught up on some paperwork. Our new med tech, Jim Brandies, was trying to learn his way around. Tom Dole and Doug Stark were nowhere in sight, not that I would have noticed anyway. We learned that morning that Bob Patton had died. I prayed for his family in sympathy and gave them my thoughtful condolences.

That morning I learned a little more about Dr. Beanie Coulters. He was sixty-two years old and had been in family practice for a while. He had served at a US base hospital in Okinawa in World War II. He had been promoted to colonel rapidly, and then, since 1947, he had been involved with nuclear medicine. He had quite a story to tell me. He had been married for thirty-seven years at that point and had recently taken up snow skiing. He had a younger brother, whom he was very close with, and an older brother and sister who lived in Kansas. He was one of the investigators of the three deaths at the Idaho Falls Nuclear Power Plant disaster. What a horror story that was to hear. He was small but husky with a graying crew cut and a matching colored beard. He came across as very well spoken in our short time together.

After a delicious lunch on board, I drove into town to fill Randy's truck with gasoline in the afternoon. I just happened to find Stan's office in the Carson area shopping center. He was off that day. Dr. Coulters bought some cigarettes, and we drove back to the ship. After dinner, Dr.

Coulters and I climbed to the rigging deck to get a nighttime view of the city's lights. I went to bed to read at 2100 hours. It was the last day of May. I was missing out on a special session of the delegates to the CMA on that day. I regretted not being in attendance, but at the same time I felt content to be just where I was. I hoped my esteemed colleagues would solve the malpractice insurance crisis that all physicians were facing. I was up at 0600 hours, had breakfast, and then set to work on a field day in the hospital.

That day I learned how to operate the suction machine, the intermittent positive pressure breathing apparatus, the anesthesia machine, the ultrasound, and the electrocardiography machines. We also cleaned and reorganized all the medicine cabinets, much to the chagrin of Tom Dole. On the other hand, Jim Brandies was a real worker. We cleaned out the hospital area, again much to Tom's chagrin. Dr. Coulters had a good rap with Tom that night. He had the same impression of Tom that I had.

We had a laceration of a foot that day. Mark Raymond stepped on a razor blade. He cut the bottom of his foot just proximal to the metatarsal-phalangeal of the right fifth toe. The wound was prepped with Phisohex; I injected it with a xylocaine for anesthesia and placed two silk sutures and two sutures of nylon. I was a little concerned that Mark might have nicked the flexor tendon in the accident. But I felt the danger of T-ing the cut to explore further might be too great an invitation to infection versus the usefulness of being able to flex the little toe.

The ship got into position for beginning the stop operation, and everything was going smoothly. We hoped to begin flooding the moon pool at 0300 hours that night, which was forty-eight hours ahead of schedule. I made it up to the steel beach for my run in the morning and then took it easy for most of the rest of the day.

Chapter XX

ONE DOOR CLOSES AND ANOTHER OPENS

On Sunday, June 1, our supervisor, Ray Benkoe, announced to me that the medical department would go on twenty-four hour duty. We were to have a live, awake body in the hospital twenty-four hours a day. I brought the message to my crew. "Okay," said Jim Brandies wryly, with a shrug of his shoulders.

"Oh, I'm sure he meant only to midnight," suggested Tom Dole hopefully.

"No, it's to be twenty-four hours. That's what he wants, and that's what he will get", I corrected.

Tom was a little sullen most of the day after that. He really did have a problem. He painted the heater in the hospital head, and Jim washed down all the decks. The place sparkled, and everything was in order. I taught myself how to use the gas machine, which checks a person's blood gases. It was a full day. Mark Raymond's foot looked good.

The CV was lowered through the gates that afternoon for wet testing. Everything was going like clockwork. It was foggy, sixty-two degrees, windy, and choppy. We never did see land during our cruise through the foggy haze.

Dr. Coulters was from upper Michigan, he told me, not far from Lake Superior and about 150 miles from Duluth. He was quite a guy. I was delighted that he was confiding in me. I could almost see the weight coming off his shoulders as he unburdened some of his own trials and worries. It was good therapy for him, I thought.

I took time to catch up on my correspondence that day at the hospital. Not only did I write the letter to Laura, but I also wrote one to General Motors, one to the clinic, and one to my good friends Vernon and Anne Freidell. What a change our world had taken since Dr. Flickenger first spoke to me about heading the medical department of this project. Vernon and I used to dive every Sunday off Isla Vista ("island view" in Spanish) near my home in Goleta for our limit of nine-inch abalone. It would take us only about half an hour after we swam out about a quarter of a mile offshore to pick our limit of five each time we went out. Abalones were so plentiful then that we would see nine-inchers growing on nine-inchers! We hadn't been diving together for abalone since this project began.

The ASK system was tested, and we found that one of the transponders from the sea trials was still beeping away on the bottom of the ocean. Also, five Soviet missile-tracking ships were rendezvousing four hundred miles north of Midway Island. We wondered if they suspected that we were on the way back again.

Mark's foot was clean with minimal swelling, but it was a bit tender still. He was also developing sore spots under his arms from pressure from his crutches. We readjusted his crutches to lessen the pressure.

I ran my laps on the steel beach that morning, but that night I felt particularly fatigued. I hoped I wasn't coming down with something. We adjusted our duty schedule to fulfill the new orders from Ray Benkoe. Jim would take the 2000 to 2400 hours duty. Tom would take it from 0000 to 0400 hours, and I would be down there at 0345 hours. I headed for bed at 2030 hours.

On Monday, June 2, I felt up to my usual exercise program. I thanked the good Lord that I wasn't coming down with some malady. I had decided that I would establish twelve-hour duty shifts for us. Jim Brandies would take 1800 to 0600 hours, Tom Dole would take 0600 to 1800 hours, and I would be available on my pager call twenty-four hours a day.

It was amazing to me how easily the crew had returned to the skill level they had last year when we were on Project Jennifer. Everything had gone smoothly that day and very professionally. The CV was back up, and they were rigging the bridle to it. The seawater was extremely clear, with visibility fifty feet or more. The hours just seemed to fly by.

I was sure glad Jim was working on the hospital inventory. We had found a lot of things I didn't even know were on board. That day the powers that be agreed to put in a voice intercom between the hospital and the office, which would prove to be a real asset. We had both the *Harold T* and the chopper out that day, so we received a lot of mail. I felt really sorry for Tom. He didn't get anything in the mail call again. I knew he was quite disturbed about things back home. I wished that I had a diagnosis and therapy regime for his depression. I also wondered if he would be able to handle a twelve-hour shift unsupervised. As a precaution, I would be putting in sixteen-hour days until I was sure he could handle it.

The CMA special session was over. All physicians would face an extra assessment on their dues. It would be an extra $60 for the AMA, $50 for the County Crisis, and $300 for the CMA. It sure added up for all doctors. The clinic would absorb it for me, but there would be a lot of doctors in private practice who were going to really get dogged by the additional fees. The Speaker of the House of Delegates for CMA, Dr. Joe Boyle, said that the legislature was going to have to give us some meaningful relief from the skyrocketing medical malpractice insurance rates by that September. I wasn't planning to hold my breath over that demand.

That night was the second night of the twenty-four-hour mandated duty hours. I would check on both the guys during their shifts in case Ray was to make a visit to the hospital or the offices and find no one awake. I wrote a letter home to Laura and then got ready for bed.

I got up early on Tuesday, June 3, to greet Tom as he went off duty. He said he saw only one patient all night. I would write a report to Ray in the medical office about coverage when these trials were over.

We were having trouble with the heave compensator that day because a sensor of some sort was out of place. The crew worked on it all day long. They wanted it fixed before they hooked up the bridle to the CV. I took time to show Jim the anesthesia machine. We also went over the defibrillator, and our sound meter arrived that day. I read up on that as well. Tom got up at about 1300 hours when we had boat drill. He was grouchy because he only had about six hours of sleep. Mark was in for a check of his foot. It was still tender, but he was back to work hanging curtains in the "ghetto." That was the name the forecastle crew had given to their quarters up forward on the ship.

That afternoon they finally began hooking up the monster-sized bridle connecting the pipe string to the CV. At about 1615 hours Dr. Coulters came to the office. He wanted to go up to the top of the derrick that afternoon. We took off and started our ascent. It was strange to be in a rolling sea and climbing up the gimbaled tower. Dr. Coulters had a tad of vertigo about halfway up and decided he was high enough. I continued up and finally topped out at the peak. I had a number-two pencil with me and wrote on one of the steel members, "Jack Rutten, MD, loves Laura Rutten—25th anniversary year, June 3, 1975." I then drew a heart around my declaration.

Dr. Coulters had waited for me on the rig deck landing, and we returned slowly to the main deck. His vertigo had disappeared by then. We then headed off for filet mignon and lobster tails for dinner. We would do hospital inventory the following day. The hospital looked great. The med techs kept it white-glove clean all the time, and I was proud of them for their work.

I was awake at 0530 hours on Wednesday, June 4, and up before my roomie, Ron Hamler, again that morning. It was gray and drizzly, but warmer than yesterday. We had undocked the CV, but we were still having problems with the lower yoke. The navy had apparently announced they were going to hold air-to-sea missile maneuvers, and would we be so kind to move our position. It was reported that our captain had answered, "No, check your priorities with Washington, and then go play somewhere else." They did, and then they did. Dr. Coulters and I made hospital rounds with our new sound meter. The hospital's office was showing about 65 decibels. The well ran at 90 decibels, and the thruster pump room ran at 104 decibels.

I then checked out our divers. They had been in the water all night going through the deployment phase for the CV. We adhered to the repetitive dive tables religiously. Would you believe our GM Rent-a-Frogs had made more than fifteen thousand working dives without a decompression accident?

I was dismayed to find Dick Dart sleeping in a chair in the decompression chamber room. Paul Santi was standing up with his head in his arms, leaning on the chamber communication shelf. Merle Garbor was sitting in a chair, still in his wetsuit, sound asleep. They had been up all night, poor guys. They were totally bushed. But the records showed that all tables were followed, and there had been no signs or symptoms of decompression sickness. Thank God for that!

Dr. Coulters came by that afternoon and indicated he'd like to take another look at the derrick. We again climbed up to the rig deck. He decided that that day was going to be the day he was going to the crown of the ship. Would I go with him? I would, and I did! He had no problems with vertigo that day. He too, wrote his name next to mine, and added, "Have a nice day!"

Inventory was on going in the hospital. I reported on the sea trials and mating that was then forwarded to Dr. Flickenger. I learned that Philip Watson, the Los Angeles County tax assessor, had set three million dollars as the cost of taxes to Summa Corporation for our operation.

I ran sixty laps on the steel beach that morning. Dr. Coulters walked with me for about a mile. After dinner that night, I checked out the crew in the hospital and then went to bed early to read my next book from the library.

On Thursday, June 5, the pipe was going down smoothly. I continued the inventory in the hospital. We sure had a lot of strange or outdated medicines on board. I didn't know what anyone would want with fifteen pounds of Epsom salts and gave it, along with outdated injectable Chloromycetin, the deep six. We had a lot of Neosporin powder, which I had second thoughts about, and also a lot of ferrous sulfate tablets. I decided to hold on to those for the time being.

I wrote a report to Dr. Flickenger listing the noise testing levels and also some of the little things that were happening in the hospital. I talked to Ray Benkoe about my perception of the questionable effectiveness of Tom on the mining leg and was referred to one of the supervisors, "Red Jack". Well, I decided to just keep quiet until I could communicate with Dr. Flickenger on that subject. The only trouble with that decision was that if the powers that be agreed with me, there were only two weeks to find someone else to fill his position. But then, such an action might stimulate some of the other med techs to shape up.

It was peculiar that I didn't get any mail the day before. The thought occurred to me that Tom might have held up my mail just to see my reaction. I could see how I must have been getting paranoid about things and then dismissed my thought.

We were more than halfway down with the pipe string that day. Moe Borncamper gave Dr. Coulters and me a good discussion about the construction of the well and the ship. Two giant A frames held the ship together in the well area. This steel was very strong, but very brittle. It had to be heated before it could be welded, and cracks tended to propagate if they started. Dr. Coulters and I looked up from the main deck to the tower, which rode so easily in the frame. The bearings of the gimbal, said Moe, were built in Germany because our country didn't have the hardware to mill them. They were constantly lubricated by an oil bath and mist. Even when the ship was in port, the bearings were required to continue to move or they would develop flat spots. The two heave compensators also must stay in sync at all times, or the tower would fall over. Moe told me of two incidents in the year before where we nearly had had that happen in the attempts to raise the TO.

There were also expansion joints on the ship that groaned and creaked but allowed for relief of the stresses on the ship. Moe said that during construction he had had to inspect the welds constantly to catch sloppy flaws left by workers. Moe then took us up to the rig deck for a more detailed inspection of the tremendous engineering capabilities and specifications needed for the pipe to support the CV.

On the morning of Friday, June 6, I took out a crewmember's sebaceous cyst at the hospital. While I was doing this procedure, Dr. Coulters came in to visit and told me about his exploits at the Nevada Test Site back in the late 1960s and early 1970s. He had seen more than three hundred nuclear shots in the Nevada desert and at sea. I listened with bated breath, but he didn't get into any juicy details. My impression was that he was quite security conscious at all times.

There was a big hold on pipe laying that had begun the previous night and continued throughout the day. It seemed the sensors on the lower yoke weren't responding properly and holding the pipe securely.

There had been sharks in the well the previous night and all that day as well. Some of the crew had caught a few with hooks baited with meat. There were also some little diving birds after the small fish. They looked like little ducks, and went to beat the band underwater, but couldn't seem to dive deep enough to get out of the well. The other visitors in our "aquarium" were giant purple and white jellyfish.

I continued with the hospital inventory that day. Jim Brandies kept us spellbound with stories of his being in the embassy in Laos and then in Saigon. He knew Dr. Lee Sanderen, who used to be in family practice in my hometown of Goleta and was a regional medical officer in Southeast Asia when Jim was there.

I received approval from Dr. Flickenger to take a holiday in July. I also dropped him a note about Tom Dole and Doug Stark. Tony Arami taught me how to start the compressors for the decompression chambers that afternoon. It wasn't nearly as difficult as one of the divers, Mike Sundberg, had led me to believe. I supposed that was really only for his own job security.

One of the major events that day was my receiving another invitation to work full time with the Department of the Army. I would have to involve Laura in some genuine consideration of this offer before I would accept it. It was to involve eleven years overseas as a regional medical officer, with subsequent full retirement benefits. There would be no hurdles to overcome with my liability insurance or job security with managed medical care.

John Wentz broke the proximal phalanx in this right fifth toe that night when he literally kicked the water cooler. It was undisplaced, but the toe was pretty swollen and black and blue. I splinted it to his fourth toe, with a little cotton between them. I wondered whether he had learned anything from the event.

The food that was served to us that day was great! Tom Dole slept ten and one half hours that day. I hoped they would get back on with laying pipe again soon as I was hoping to be home the following weekend.

Saturday, June 7, dawned with fog and drizzle again. We bottomed out with the pipe string at 0400 hours. We were to spend the next forty-eight hours testing the CV. Then we would pick up the pipe and head for home. I was told the "dildos" on the lower yoke were remanufactured and had worked splendidly!

I held down the fort at the hospital while Tom slept and Jim inventoried the storeroom below. We had an inspection of our facility followed by a tour with the captain later that day. Dr. Coulters went with us on the tour and got a real eyeful of this ship. We had our pictures taken, and they turned out just fine. When we got to the ghetto, Jim Danzen remarked about John Wentz breaking his toe the night before. He asked one of the crew standing there, "Was he chewing gum?"

"Yeah, as a matter of fact, he was!"

"I might have known. He never could walk and chew gum at the same time."

Later, when Jim Danzen was referring to John Wentz and Mark Raymond he called them "Grace I and Grace II."

There was a big school of porpoise off the fantail that morning. There must have been a hundred! They played for an hour or so.

Dr. Coulters began to tell me a story about this wife, Dottie. A year ago she had a rectal polyp. It had turned into cancer. She had a combined abdominal-perineal resection with a permanent colostomy. Then she had developed adhesions. She underwent surgery again. She had subsequently developed a tremendous lymph edema. We were interrupted when someone came into the cubicle and the conversation ended. I thought he was really in the process of undergoing a therapeutic verbal catharsis with me. I hoped later he would open up to me again. I thought he was holding back a deep animosity toward someone. I sensed there was a lot of grief there that had not yet been purged.

We watched the television cameras on the CV do their thing in the control center that afternoon. It was fantastic to watch. It was as clear as though it were filmed in air instead of water, almost.

Tom told me later that afternoon of he and his wife's upcoming wedding anniversary. His great-grandfather was named Doty. He told us he had been shot by Comstock in the Montana Anaconda area as a claim jumper. He said his current job was the first really good job he had ever had. I hoped I hadn't done him in with my recent report back to Dr. Flickenger.

I felt very fatigued that night. I decided to go to bed early and read. I forgot to mention that at the conclusion of the inspection that day, we heard from the bridge the faint sounds of a bagpipe wailing. Following the noise, we crossed to the port side of the forward bridge. The music was coming from somewhere below. We descended to the main deck, continuing to hear the unmistakable solo. The sound was coming from the area of the paint locker room. Opening the door, we were confronted by Honeywell's Jerry Wall playing a regimental march on his pipes. He was very good too! Later that evening we watched the film of the recovery/exploit from a year ago. That was good too!

After breakfast on Sunday, June 8, Dr. Coulters and I walked up to the steel beach. He finished the story of his wife Dottie's recent health setbacks. I was relieved to learn that it hadn't been what I feared most. It had a happy ending. She had finally licked the colostomy thing. She could eat whatever she wanted. He said she did tire easily, though. They sure sounded like a lovely couple. Dr. Coulters also told me he had had surgery for a malignancy, a seminoma of the testis, in 1954. He said he'd had his nodes stripped all the way to his diaphragm and was given two thousand rads to his gonads as a precaution. He had subsequently developed a bleeding peptic ulcer. He had an 80 percent gastrectomy also.

That day he and Jim completed their oxygen tolerance test. They both had done very well with it. They were still holding the CV on the bottom. We could see on the monitors that everything was going just fine. My med techs and I completed our inventory that day, and I did the summary of the diving activity on the mating leg.

At 1600 hours I went to the steel beach and ran my laps with Dr. Coulters. My knees gave out on me that afternoon. Even though I changed direction on each mile, my knees just couldn't take the pounding on the steel deck, even though it had that indoor/outdoor carpet on it. They took movies of us in the surgery unit that day. I saw myself and decided I needed to lose weight and then checked to see that I had reached the heaviest I had ever weighed at 183 pounds. That night was a difficult night sleeping due to miserable pain in both my knees. When I awakened on Monday, June 9, my right knee was really swollen.

They were pulling up pipe at a great rate when I got out on deck. I thought that at that rate we would be back to port the following day.

I checked over our final inventory sheet again and ordered the last of the things we needed for the recovery mission. Dr. Coulters and I spent over an hour in the pipe well watching the activity there. I am unable to relate to you my estimates of the size, weight, or number of pipes that were being reassembled in the well, but it was really mindboggling!

Poor Tom, he hadn't autoclaved the instruments I had set out for him to do the previous night. To make things worse, he overslept by one and a half hours for the second shift in a row. Jim Brandies told him that morning he didn't feel sorry for him at all. There didn't seem to be any reason his wife was way up in Seattle. She could easily have gotten a job down in Southern California if she had really wanted to.

The CV was back in the well on Tuesday morning. The gates would be closed at 2100 hours, and we were to begin pumping out the well. I was already anticipating the ship's bucking like a bronco when the gates were shut. We planned to make a stop back at Catalina Island and then head for home.

Tom met me on the boat deck that morning. He was bound and determined that he was going to be a physician some day. We stood and watched the docking legs bring the CV up. The GM divers presented me with a Styrofoam cup that had been sent down to six thousand feet with the pipe. It had been compressed to thimble size. It was a great souvenir and a visual reminder to me of the great stresses that came to bear when encountering depths of that magnitude. They had proudly written "Delco Rent-a-Frogs" on it, and I kept it for many years in a hallowed place on my shelf at home.

Doug Stark came out on the helicopter that afternoon. He told me he didn't think Tom was qualified for the mission either. Later the chopper brought out more people, including Bill Kelly, who was the person who had refused me permission to view some of the secreted remains that were brought up from the TO last year. Another guest was six-foot-four-inch Erwin Rast.

There was a very detailed briefing held later that afternoon. Dr. Coulters was invited to it. Bill Kelly told us that Dr. Borden was coming down at the end of the month. I suddenly had the feeling that I was on the outside looking in again. I was astonished that Bill just didn't look the same to me as he had the year before.

Luis Parent awakened me on the morning of Wednesday, June 11, at 0530 hours when he opened my stateroom door and complained of an earache. I told him I would come down to the sickbay and take a look as soon as I got dressed. When I got there, he told me he'd seen Tom earlier that morning. Tom had tried to dig it out, but couldn't remove the plug. While I was getting dressed, Jim Brandies had a whack at it and irrigated it clear. It just happened to be Luis's birthday that day, and now he was feeling no pain.

We had gotten underway at about 0000 hours that night. We had done some steering tests and cleaned out some transponders on the way home. Our estimated time of arrival was 1300 hours on Thursday, June 12. There was a feeling of nostalgia evident in the crew with this floating museum. I wouldn't be sailing on her again, unless the A Crew brought home a new portion of the TO.

I took diver Dick Holding to the Avis Car Rental, Dr. Coulters to the airport, and Al to McDonald's in Santa Barbara, where he met his family who lived over the mountains in Santa Ynez about thirty minutes away. I then made it home at 1815 hours, which was just about on schedule.

On arrival I was entertained by my family and our menagerie of dogs and cats, along with a good two-olive martini and filet mignon barbequed on my own outdoor grill. Raul had just gotten home from his high school graduation and was leaving that evening for the party for seniors' night at Disneyland. Randy would be home much later, as he was working delivering pizza at the nearby university campus. It sure felt great to be home again.

On Friday, June 13, my first full day back home, I was awakened at 0630 hours by a phone call from Dr. Flickenger. I was told that there was to be no mission. I expressed my disappointment at the news but affirmed to him that I was truly interested in a career with the government as a regional medical officer. He said he would look into that for me. So much for my plans for Jennifer, Azorian, and Matador missions, it would seem.

Nonetheless there was no dearth of tasks for me that day with all I needed to catch up on. There was also a nice welcome back to the dispensary at General Motor's plant from seven to nine that morning. It didn't take me long to get back into the routine of work sans DOMP. My report to Dr. Flickenger on diver activity during mating operations was finished by Monday, June 9, and I got it off to him.

On Thursday, June 19, Dr. Flickenger telephoned again. The Department of the Army needed a regional medical officer overseas. They were willing to add on my navy duty during World War II and the Korean conflict, and if I would commit to eleven years overseas with them, they would retire me with a Civil Service retirement. He made an appointment to meet me at the PO in Los Angeles on Thursday, June 26, to discuss it in a secure venue.

I quickly relayed the information to Laura. What a teammate! Laura said without hesitation, "It sounds exciting!" We agreed that the kids didn't really need us to be there at home anymore. Randy was going to be in law school that next year after finishing his studies at the University of California-Riverside, and Raul was to be in college in Santa Barbara for the next two years, and then likely he would be headed overseas as well in the commercial diving industry. We both thought it would be an exciting and adventuresome new career for us.

On Thursday, June 26, Dr. Flickenger and I met at the PO. He gave me all the papers required for the government application. The information also included requirements for verification of my navy duty in both wars. We shook hands on it. He would keep me informed of the progress of my application.

I was called back to the PO on Wednesday, July 2, for a conference relating to the lawsuit of John Mackel. I invited my mother to come along with me, and after the conference, I took her to tour the Queen Mary in Long Beach harbor. She was just delighted with our adventure.

I offered what information I could give to the conference regarding the unfortunate risk John had taken when he hurried to catch the flight to San Francisco. I did think we had good evidence to present that he had healed his previous damage but that the etiologic agent—arteriosclerosis of the coronary arteries—had persisted and had caused his death.

July was an otherwise routine month of my usual practice, medical society meetings, and community service. Laura and I did take a Mediterranean cruise from July 9-23. The cuisine was of the quality of the *HGE*, but the accommodations were not much more luxurious than the Spartan quarters of the ship.

On Tuesday, August 12, I returned to the PO again for a meeting. It was to be demating the CV back into the *HMB-1* during the last two weeks of August. I thought all of us were disappointed that the mission wasn't going to go forward. Dr. Borden's attitude was not enthusiastic. I hadn't seen much of Jim since he left his things in my locker back in Lahaina. I guessed he had a lot on his mind. Our scheduled date of departure was Tuesday, August 19. That was the same day that Raul would be returning back to Laura's brothers' farming operations in North Dakota for a working holiday.

I slept on the *HGE* on the night of Tuesday, August 12. Dr. Flickenger held a nice dinner party for the medical group, and I decided it wouldn't be wise to drive all the way back home that night. My old bunk felt like home, even if I did miss the closeness of Laura.

Chapter XXI

ONE LAST VOYAGE

I got up at 0530 hours on Wednesday, August 13, showered, shaved, made my bunk, and headed for the mess hall. Dr. Coulters and Dr. Charley Backes from Washington, DC, were already having their first cup of coffee. I had my half grapefruit and coffee with them while we discussed the upcoming mission. Dr. Coulter was his usual affable self. Dr. Backes, whom I had never met, seemed reserved. He did seem to know Dr. Flickenger very well, and I was sure that I would find out what part of the mission he was to be involved with when the time was right.

Dr. Backes and Dr. Coulters were due to fly back to Washington, DC, later that morning. They were delighted by my offer to drive them to LAX. It was only an hour and twenty minutes to home after I dropped them off at the departure gates.

On Tuesday, August 19, my son Raul and I were up at 0515 hours. He had to be on board his Northwest Orient flight at 1110 hours that morning headed for North Dakota for his working vacation. He was going to be eighteen years old in three months, so he was going through the same doubts and hesitations about leaving home that I had when I was going into the navy a week after my eighteenth birthday. I could empathize with him—but he had a ticket back. My departure had been open ended.

It was interesting to watch him as I sat with him in the flight waiting area. It was a big DC-10, and he was to fly into Minneapolis. Laura's nephew would be meeting him there to take him to Langdon near the Canadian border. His boarding call was made, but he continued to sit with me. Finally, all the passengers were on board, and the flight attendant was looking at her watch. "Aren't you going to go?" I asked. He rose without speaking, gave me a big hug and kiss, handed his ticket to the pretty, young lady, and got on board.

I drove to Long Beach and the *HGE*, arriving in time for lunch. I greeted a lot of old friends, had a bowl of split pea soup with ham that was delicious, and found the flounder stuffed with crab very tasty as well.

I made a call to Laura to let her know that Raul had gotten off without incident and that we were going to leave at 1000 hours tomorrow. We had been planning to leave that night, but the Los Angeles County tax assessor, Philip Watson, had made another political play. He stated that he wanted proof that Summa Corporation did not own the ship. He had a court order staying our sailing. It was finally undone by the justice department, but our departure was delayed.

Dr. Coulters was back on board and had brought a number of people with him from the testing lab in the bay area. They were to do a survey of the ship while we did our thing. It was good to see all the folks again. Tom Dole and Jim Barlough were a little uptight from all the waiting. My old roomy, Carl Atkinson, was his usual droll self. Captain Tellaman would be skipper for this voyage.

After dinner I got all my things put away. I was ready to retire about 2300 hours. I was sorry that Laura was all alone that night. I thought I had better remember to do something nice for her tomorrow. A dozen roses might be just the right thing. I telephoned directory assistance to get the number for a florist and decided to call them in the morning to have flowers delivered for our twenty-fifth wedding anniversary.

My first thought when I awakened the next morning was, "Happy silver anniversary, Honey!" There had been a full moon and clear sky the night before. It would have been a marvelous evening to sit on the patio by our pool with my bride and reminisce about the honeymoon we had twenty-five years ago. She was only nineteen then, and I was twenty-three. We had a friend who had a cabin on an island in Lake of the Woods just out of Kenora, Ontario. He had offered it to us, with all the amenities, for a whole week.

We actually had spent our honeymoon night in a Victorian hotel in Winnipeg absolutely exhausted from the festivities. The next day we drove on to Kenora, chartered a launch that took us to the cabin with our luggage, and dropped us off. The cabin was well stocked with the necessities and had a marvelous view of the lake. We explored the island and the area all around our cabin. The boat was stored under the floor that was raised on pilings. We put it into the water to soak because it was quite dried out, and it leaked.

We enjoyed a delightful dinner on the porch, but the sky was clouding up and the wind made it very cold. It was fun to retire under the down comforter on the big bed. In the morning we had snow on the ground and an icy gale was blowing a mixture of sleet and snow almost horizontally. We had to abandon the cabin and the leaky boat and move into the Kenricia Hotel on the Kenora waterfront. At least we had been together and warm. Twenty-five years later to the day, I was still warm, but we were not together.

There was not too much of a sea, and the television and radio news broadcasts said that we were on our way to the isthmus at Catalina Island. We really were not. Our course was between Catalina Island and the island of San Clemente. We left Pier E about half an hour late, 1030 hours on Wednesday, August 20. It was an uneventful departure. We had clearance of the stay on our leaving until a hearing by the justice department on Thursday, September 18.

Philip Watson would get the word on that day as to who really owned the vessel. He could decide what to do with his $7.5 million tax bill then. I understood that there was even a $3 charge for filing of the papers that had been added to the $7.5 million bucks he was seeking.

Dr. Coulters worked all that day with his people from the testing lab. They were surveying the vessel and cleaning out some filters. Apparently these were extraordinarily efficient filters, but they did contain asbestos, and so they were not being manufactured anymore. They apparently would be returned to the testing lab. I found the sick bay to be pretty messy that morning. The door remained closed most of the day, and it was really rather inaccessible to the personnel. We would change that the following day. The morale of the crew was down but would probably pick up now that we were at sea.

There was a Soviet ship, the *AGI Sarachev*, headed toward us. It was to be in our area at 1130 hours the following day. This added to the crew's anxiety level considerably.

A rather heavy marine layer descended upon us at dinnertime. I had a good talk with Carl that evening. He was sitting at his desk in the stateroom having a couple of beers, but he seemed to be a bit jovial and friendly. I sat for a while watching the news, but it seemed to be all bad news that day. I did see a few pictures of the ship leaving the harbor though. I thought that we would be out of range of the television news the following day, so I made myself watch it so that I would be caught up on the events. I did manage to get to telephone the local florist shop back home before we left port and get a dozen roses on the way to Laura for our special day.

Thursday, August 21, dawned gray and overcast with light seas but a rather long swell that gushed past the bow with a soft sound. The Soviet ship was to be in our area that day, but with our radar set at a forty-eight-mile radius, it was not detected.

The sick bay was clean, and the doors were opened before 0700 hours. The med techs didn't seem to mind the change in the plan of the day at all. Tom slept most of the morning. Dr. Coulters was still busy with the things that his people were doing. The ship was clean as a hound's tooth.

Jim Barlough broke out the cribbage board that he'd stowed for a long time, and I refreshed my memory at playing cribbage with him. I also read half the symptoms on dysbaric osteonecrosis and listened to some of my audio digest tapes.

The ship's course changes were fun to watch from the after bridge. There were no other ships in sight and no helicopters either. I figured that we were probably out of their range, so there would be no television for the same reason.

I took in the movie, *The Bedford Incident,* in which a US Navy destroyer and a Russian submarine play cat and mouse until the final moments when they simultaneously destroy each other. It was enough to give one the chills.

Nothing else of great interest happened the rest of the day. There were no illnesses or injuries to report. I hoped that Raul had a good trip and that he had reported to his mom. I also hoped that Randy had gotten home all right.

On Friday, August 22, it was overcast again, and we were still running a course in squares. There was an unidentified vessel running parallel to our course just off to our starboard side about twelve miles distant. When we would turn toward her to close the distance and get a look at her, she turned away. Gradually, she pulled away until our radar could no longer track her at forty-eight miles.

We continued in our square pattern. Then, at about 1030 hours, we turned to sixty-eight degrees and headed for Catalina Island. By dinnertime we had San Clemente Island off our starboard bow, and Catalina Island was just off the horizon. A smaller island was seen to the north earlier in the afternoon. It must have been San Miguel Island.

The day lightened up gradually, and in mid-afternoon the sun was shining through a broken sky—but not very warmly. Jim Barlough and I played cribbage and read through the day in the sick bay, available for the walking wounded if they needed attention. When the troops were interested and involved, there was little time to be sick. Crew morale had skyrocketed since we had been underway. I even took a nap that afternoon.

The thick fog had rolled in as it typically did at this time of the year along the California coast. It was a very dark night again. We planned to be off the isthmus that evening and hoped to sneak back home.

Dr. Coulters and friends had a good, long discussion about sea-stories in the mess hall. I didn't join in, since it seemed to be a "family" thing with them. We finally got in range of mainland television again, and I watched the Los Angeles-based ABC affiliate news. Nothing much new had happened since we had been out of touch, but there was a story out that the *HGE* had disappeared!

The August 22 report said, "The ship sailed yesterday morning for Catalina," even though we had sailed on the twentieth. "Since that time no one has seen that six-hundred-foot vessel." It went on to say that all maritime agencies, including the company that designed the ship, said they had no idea where the vessel is. The Coast Guard, which had issued a notice to mariners that the ship would be conducting tests off the island, said it had no idea where the ship had gone. The navy, the Marine Exchange, and the Office of the Port Captain in Long Beach where the vessel docked all confessed that they really didn't know the ship's whereabouts.

The Saturday, August 23, issue of the Long Beach *Independent Press-Telegram's* late news final edition had much the same story on the front page, with an inch-and-a-half-high headline: "GLOMAR VANISHES ON CATALINA TRIP."

On Sunday, August 24, I was up at my usual time. We were still in the solid soup as far as the weather was concerned. We could see the outline of the isthmus, though, and the day promised to be nice later on. Numerous boats of all kinds were examining both the *HGE* and the *HMB-1* through the fog. Rowboats, rubber boats, Boston Whalers, skiffs, and yachts of all kinds, including both those of the middle- and upper-class citizenry, were among those represented.

We dropped anchors, but there was a very strong current that was running from the north past the isthmus, and it did catch a few of those who were out in the rowboats. The news still carried stories that we were missing at sea.

The *HMB-1* began to fill about 1300 hours. It would take all night to get it down to fifty-three feet, and then the divers would inspect to make sure that everything was going well before it was sunk completely.

In the meantime, we had several helicopter trips and mail call. I had three letters from Laura and one from my mom. Laura was delighted to get the roses and wrote sweet things that recharged my batteries. I also learned that Raul had arrived in North Dakota in great shape and was having a ball learning the art of farming in the land of durum wheat, brewing barley, and soy. Randy was down in Orange County area painting a good friend's home. Laura was alone at home with our pets and holding down the home fort. In twenty-five years of wedded bliss, I believed we could communicate our unwritten message by telepathy.

The rest of my afternoon was dedicated to routine medical care with no serious problems. I got nearly an hour on the steel beach and took in the afternoon sunshine. I hoped that I could get more time up there the following day. "Red Jack," one of the supervisors on board, left the project on the last afternoon chopper that day. I understood that he'd been with the program since 1969.

Monday, August 25, dawned sunny and clear with a small breeze and a slight swell. It promised to be a gorgeous day. There was a fog bank lying on the other side of the isthmus, but it didn't come across into the harbor.

Once again, there were a lot of boats around the *HGE* when I got up on deck. The *HMB-1* was slowly sinking and was then down to nearly the fifty-three foot mark.

The GM Rent-a-Frogs were working back and forth setting out markers for the mating. Dr. Coulters was riding with them in a boat. So far we had not had any significant injuries or illnesses in the sick bay that day. We did take an x-ray of the kidney, ureters, and bladder on one crewman with vague abdominal distress in the right upper quadrant. I x-rayed another man who had sprained his ankle. Both films showed no evidence of pathology.

By late afternoon the *HMB-1* was totally submerged. Everything had gone according to plan. The next day we would move over her and begin demating. I watched the news after dinner but didn't watch the movie. I read instead.

Tom was on his usual watch, smoking his cigar, and acting very secretive, as a good agent should do. Carl smoked about ten cigarettes in the two hours that we were in the room together reading. He sounded terrible and probably was plugging his bronchi to the brim.

On the morning of Monday, August 25, there was considerable positioning and determination of the ability of the *Happy Hooker* to place the anchors. The *Pacific Gemini* and the *Pacific Saturn* assisted, and the work began. The forward port anchor was first placed, and then the forward starboard. Then the aft anchors were set in place. The morning was overcast until about 1200 hours, and then the clouds broke up into a rather warm day. At 1400 hours the wind came up again and was very strong, blowing over the isthmus at about forty miles per hour. Dr. Coulters and I climbed up to the crown before the wind began to blow and surveyed the harbor from there. One could see clear across the one-hundred-yard-wide isthmus from this raised position. Our names had been washed away from the crown, so we replaced them by scratching them in with a knife. It was a great view all around. The wind was so strong that the pilot had to land the helicopter backward on the deck. The divers announced that everything on the *HMB-1* looked to be just fine.

I went into the hospital and saw a crewman who had a one-centimeter mass between his scrotum and his anus that felt like a tumor of the bulbous urethra. His rectal exam was negative. Another crewman followed him with a vague distress in the right upper quadrant that I thought was probably nothing more than domestic tensions at home.

Carl continued to smoke his pack of cigarettes after going to bed, but I guessed I wouldn't say anything to him more than suggest that he quit because of his peripheral vascular disease.

At about 2200 hours the anchor winch was working slowly, and it seemed that our aft anchors were dragging. The divers would be on that in the morning. If we dragged our anchor over our own umbilical, we would be here for quite awhile. In any event, we were committed to an extra day or so by then.

I got up at 0500 hours on Tuesday, August 26, to listen to a family-practice audio-digest tape or two before breakfast. It seemed that during the night, as they pulled up on the starboard aft anchor line, it kept coming in, and so the divers had to go down again that morning to see what was happening. Examination of the line revealed that the anchor had not turned over when it was dropped, and the flukes were pointing away from the ship. The anchor had just been dragging in the sandy bottom, and there was some concern that the ship may cut the umbilical to the *Ore Quest*.

The *Happy Hooker* came in and took the anchor out for reimplantation. We would lose another day on our schedule overcoming this hurdle. The remainder of the day was fairly quiet as we waited. Positioning was accomplished, and finally the docking legs were lowered during the night. The CV was deposited in the *HMB-1*, and all seemed to be well. There was only a three-inch clearance for the machine to enter the *HMB-1*.

Tom was showing off his little brass cannon, which was indeed beautiful. He stayed up most of the day and was up most of the night working on it. I didn't think he'd had more than about two hours of sleep in the last twenty-four hours. It was laundry day, and mine came back in good condition.

On awakening on Wednesday, August 27, the docking legs were almost back up to where they were supposed to be. The divers were in the water clearing cameras and lights and adjusting valves. Most of them were "saturated"; that is, they had absorbed as much nitrogen as they could without requiring decompression on ascent. We tried to avoid that in all their tasks. There was some talk that I might get in a spot dive or two to avoid putting any of the divers into a decompression situation. I just had to be patient until all the crew's chores were done. They still had to close the roof of the well before we could get underway.

Dr. Coulters and I played a little cribbage, and then the phone rang. It was a call for an emergency. A civilian on the isthmus anchorages had been stung by a bee and was reported to be unconscious. I grabbed the emergency and resuscitation kits and headed out with a man in his old Boston Whaler that had been dispatched to the ship. He introduced himself as we powered over to the anchorage. The patient and his wife were apparently on vacation.

The Los Angeles County lifeguards were there with a cuff on the patient, and they were monitoring him. "He's been stung on the left hand by a bee," said his wife. "He pulled out the stinger and began to swab the deck. This was at about 0915 hours. Shortly after, he complained to me that he couldn't see and then collapsed. I asked our neighbor to call for the lifeguard. The lifeguards said they could have him to Avalon in twenty-eight minutes." She was a marvelous, unruffled historian and the kind every healthcare provider needs at a time like that.

The patient was by then conscious and appeared rational. He was covered with a sleeping bag and lying in the after cockpit of his boat. His pulse was rapid, but regular and steady. He told me he had no previous history of heart problems. His doctor had recommended he have a two-week holiday, though, because he had been working too hard.

He denied any sensitivity to medications and particularly to bees in the past. His blood pressure was 118/68, with his pulse at 112/minute. I gave him half a shot of epinephrine subcutaneously in the right upper lateral arm at 0945 hours. He was shivering with marked, wracking chills, in spite of the sleeping bag covering him. His skin was clear but pale. There was no evidence of respiratory distress or cyanosis at that point.

At 1000 hours his blood pressure was 132/80 and the pulse was 112/minute. The quality of the pulse was much better, and his carotid pulses were visible. At 1015 hours, his blood pressure was 132/76 and his pulse was 90/minute. He was moved from the shade of the cockpit to the sunny area of the deck at his request. At 1030 hours he drank some grapefruit juice from his wife; his blood pressure was 126/80, and his pulse was 100/minute and regular.

He complained of some pain in his right hip and stated that was not a new problem. Apparently, he was a supplier to the *HGE*. At 1040 hours he sat up, and his blood pressure was recorded at 138/80 with pulse at 98/minute. He didn't feel very well sitting up though, and he laid back down again. At 1053 hours his blood pressure was 122/78 and his pulse was 100/minute. His heart sounds were normal, he had sinus rhythm, and I heard no murmurs. His carotid pulses continued visible and of good quality. He had no respiratory problems and no cyanosis.

At 1105 hours his blood pressure was 127/72 and the pulse was 96/minute. He seemed stable. I advised them to start for home as soon as practical. They were two hours from home and his doctor lived near the place where they kept their boat and could be reached en route.

At 1120 hours the remainder of the epinephrine was still in the syringe with a needle guard over it. I placed it in a paper envelope and gave it to the patient's wife. I gave her instructions on how to inject him if he should have another episode of syncope on their sail home. She offered to pay me, but I refused it. Their thanks were quite profuse and accepted as payment in full.

By 1600 hours we were underway for Pier E in Long Beach. There were good-byes to say to Dr. Coulters and the rest of the crew. I praised my med techs and told them I would write nice things about them when I sent the report of the demating operation to Dr. Flickenger. I had to wonder, now that we had put the CV back in the *HMB-1* and sent it back to Vallejo in the Bay Area, if we would ever sail with this historical monument again.

In any event, I promised myself that when all this was declassified, I was going to write a book about my experiences with it. I thought of this project as I had the Sea Spider Project. That project had occurred about the same time that Neil Armstrong stepped onto the moon. Both events were classic moments in history. The first manned landing on the moon had been written about at length, but perhaps my next book would be about the lesser-known project.

We were in the outer harbor about 0400 hours on Thursday, August 28. There was about a two-hour wait for dawn and the docking at Pier E. I left a few of my things in the locker just in case (and hoping) I came back soon. I knew that I would have the opportunity to retrieve them if the ship sailed for mothballs in Suisun Harbor near Vallejo.

It was just 10:30 a.m. when I drove into our driveway at home. It was a surprise reunion. The anniversary roses were rewarded with a big hug and kiss. There was a lot of mail to sort through and, of course, bills that needed to be paid. I checked back in at the clinic that afternoon and learned that there would be another special session of the California Medical Association coming up in two weeks and that I would be expected to attend.

On Monday, September 8, I left at 0715 hours for the *HGE*. Dr. Flickenger was going to have a final briefing and wanted me to write the summary of the demating operation for him. It was a nine-paragraph summary of the August 19-28 activities on the *HGE*.

My first paragraph dealt with the fact that the initial seventy-two hours had been spent in "operations not involving the *HMB-1*." This, of course, referred to the "hide and seek" with the Soviet *AGI Sarachev*. I noted that fifty-nine patients had been seen in the hospital medical office during that time.

My sixth paragraph related to the bee-sting patient and was a summary of the description I wrote above. The rest of the paragraphs dealt with sound-level testing in various compartments, instructional exercises with the med techs, and the injuries requiring x-ray or suturing. I presented the completed report to Dr. Flickenger at the party he hosted for the medical crew that evening.

I slept well in my bunk above Carl that night. Dr. Flickenger suggested that the *HGE* might be going to Pearl Harbor, Honolulu, for dry dock. I left the ship at 1300 hours on Tuesday, September 9.

The following day I was called into the administration office at the clinic and told that a Department of Defense investigator had interviewed both the director and the administrator. No doubt that had been secondary to the application I filled out for Dr. Flickenger regarding my change of career from the clinic to the Department of Defense. I was surprised at how fast that process had gone.

On Monday, September 22, I went down to the *HGE* again at Dr. Flickenger's request. I left at 0650 hours and was in the parking lot at Pier E at 0900 hours. Dr. Borden and I had orders to proceed on a long round-trip exercise that day from LAX to the Bay Area. The purpose was to be checked out for any exposure we might have received during the project.

The examination and testing took about four hours. No medications or invasive procedures were used. The only mandated instruction was that we empty our urinary bladder before we began the testing procedures. Once we were on the examining bed, we were not allowed to move. We were allowed to sleep if we wished, but the body had to remain quiet beneath the scanning machine.

The venue of the test was a pit in a laboratory that was shielded from any chance of extraneous radiation. Two of my med techs were there too. By noon on Tuesday, September 23, I was on a flight back home. I was later informed that the testing had showed no evidence of internal radioactivity in me.

Chapter XXII

DEBRIEF AND WALKING THROUGH THE OPEN DOOR

On Wednesday, October 1, I drove down to Pier E and the *HGE* at the request of the legal counsel for John Mackel. They were taking depositions from those of us in the medical department who were involved at all with John and his illness. My only activity with the case was that, without seeing the patient and just on the basis of Dr. Borden's reports from the PO, I did not justify medical evacuation of John immediately after the incident. They took my deposition after they had taken those of the med techs.

I had lunch on board and had a chance to talk with the A Crew's chief steward, Ron Hamler. Ron told me that the crew was down to only sixty-eight members, so he didn't have much to do for mealtimes. He was still turning out quality chow with more than one entrée option though.

On Monday, October 6, I flew to Chicago to be awarded a Fellowship in the American Academy of Family Physicians. It was an impressive experience, but it didn't carry any additional remuneration. It only added the initials FAAFP behind my medical degree.

On Friday, October 10, I received a call from a Mr. Glenn Steele in Washington, DC, asking me if I could come to his office on Monday of the next week for four days of indoctrination and physical examination for the new job with the Department of the Army. Laura and I were both excited about the probability of a new career in government service. I told him I would be there on Sunday, October 12. The clinic had been very understanding about my repeated absences from CHEC. By then the staff was so efficient that they really didn't need me, except for the occasional times the computer would crash.

That Sunday I left at 6:50 a.m. for Washington, DC. On board with me was my good old diving buddy and colleague, Vernon Freidell, MD. He was going to Philadelphia for a meeting. It made for good company on the journey and settled the butterflies

I had regarding the implications of a new job and my uncertainties of what my duties would be. Just after we passed Columbus, Ohio, we suddenly noticed smoke coming from the overhead bin of the seats two rows in front of us. A flight attendant arrived almost at once, shooed the occupants out of the seats beneath the smoking bin, and with a carbon dioxide extinguisher, flooded the bin with a thick, white cloud. We saw no flames, but there was still something burning. The attendant set down the extinguisher and ripped the overhead light fixture out of the ceiling. Apparently, there was a short there. This event was over within less than a minute. There was no panic on the plane. The pilot came on the intercom and told us that there was no emergency and that we would continue our approach into Dulles Airport. We arrived on schedule, and Vernon hurried off to make his connection to Philadelphia.

I headed for the baggage claim and Glenn Steele. My bag was intact, and Glenn was recognizable at once from the description that was given to me over the phone. He had a room reserved for me at the Holiday Inn in Roslyn, Virginia. As we drove the twenty-five miles in from Dulles, he briefed me on the agenda for both my physical and psychological examinations. He invited me to his home in Annandale, Virginia, that evening for a barbeque. He told me that Dr. Flickenger would also be there. The following day would be a day for me to sightsee the capitol and the museums on the mall, and I would be on my own. The next three days would be packed with must-accomplish items that would leave me almost no time for social events.

Glenn was a pleasant young man in his early thirties, about the same size and weight as I, with reddish hair. It was obvious that he'd welcomed newcomers to the "factory" before. He had everything planned to the minute. After the barbeque that night, he told me I should eat sparingly, avoiding red meat, alcohol, etc., so that Tuesday when

I appeared at the doctors' offices, I would be fasting for blood tests.

Glenn's wife, Willa, was a vivacious, dark-haired woman with a personality that smiled. Their two sons, Sid and Scott, were nearing their teens. Their home was in a deciduous forest with leaves turning to the varied shades of autumn. After dinner Glenn gave me three phone numbers that I could use during the week to contact him, should there be a need. He drove me back to the hotel at about 10:30 p.m. It had been a very long day for me, and I was bushed.

On Monday, October 13, I awakened to a beautiful fall day. I had a humble breakfast in the upstairs dining room of the hotel. There wasn't anything on the menu that looked anything like what I was used to eating on the *HGE*. I toured the Arlington Cemetery, the Kennedy brothers' burial sites, the Iwo Jima monument, and then crossed the bridge to find myself in the mall at the Lincoln Memorial. The day passed quickly with so much to see and enjoy. I even got as far as Union Station before I headed back to my hotel.

I eschewed breakfast on Tuesday, since I was to be fasting for my physical examination. Glenn had told me about the shuttle bus that stopped at the building right across the street from the hotel. I was to identify myself to the security guard at the desk. After he verified my appointment by phone, he handed me a chit for the bus.

Glenn met me in the lobby of an office building, where I was given a temporary badge by security. I followed Glenn to where I was plugged into their system for the day. It was the usual laboratory-screening thing, with a dozen history and psychological evaluation forms to fill out. At noon there was still a lot of paperwork to do.

Dr. Flickenger appeared at my last station and invited me to come with him to the VIP dining room for lunch. He introduced me to a number of his colleagues in the dining room. The food was delicious, but the ambience was a little too Victorian for my taste.

The day finally ended. I was told to take the shuttle again the following morning back to the office building to undergo a hands-on physical examination and several psychological interviews.

After a thorough examination the following morning, apparently there were no glitches. The technicians seemed pleased with the quality of the testing. As I left Glenn told me that my plane would leave at noon the following day. He reimbursed me for my hotel and flight fare, and we said our good-byes. I had dinner back at the hotel and retired to my room to watch the second game of the World Series between the Reds and the White Sox.

The next morning I took a cab from the hotel to Dulles airport. The plane left right at noon, and I was home by 4:30 p.m. I received another warm welcome at home and answered at least a thousand questions before heading off to a much-needed rest.

On Friday, October 17, Dr. Flickenger greeted me with a happy phone call. He was delighted that everything had gone as well as it had with my trip back to Washington, DC. He advised me to ask for admission to the job on a level GS-15, and I did. I later learned that was to be my entry level for my new position. It was to be a great adventure and one that I was anxious to take on.

I got back into my usual routines at home and got down to visit Tom Dole and the *HGE* each Wednesday. There was not much action for me there with only sixty-eight crew members on hand. The crew recycled the machinery on a regular schedule to assure that everything remained viable.

Then on Wednesday, October 29, the "praying mantis" dropped a section of pipe. It was a near disaster. Had the carriage come down the ramp, the pipe would have pierced the deck just above the hospital. On Wednesday, November 5, Tom finally got his raise.

On Wednesday, November 12, I picked up Dr. Flickenger at the PO and brought him to the *HGE*. Dr. Baxlow's contract was to end the following Sunday, and he wanted to thank him personally for his assistance with the project. During our visit on Wednesday, November 26, we met with the lawyers who were still working on several of the medical cases that were in litigation.

The *HGE* was still at Pier E through December, and I continued to visit the ship each Wednesday over that period of time. It was a bit like going to the wake of a dead friend for me.

On Tuesday, January 27, 1976, Dr. Flickenger called to tell me that the *HGE* would head to mothballs in Suisun Bay on March 31. He also told me that Harvey had been permanently closed on Thursday, January 8. On my visit to the *HGE* on Wednesday, February 4, I was told that the CV had been cut up and salvaged for $144,000. On Thursday, February 19, I had my last visit to this marvel of a ship. It felt like I was visiting someone's tomb.

A telephone call from Glenn Steele told me that the first overseas assignment for Laura and I would be to Tehran, Iran, and that my duties would be as regional medical officer for the subcontinent: Iran, Pakistan, Afghanistan, Nepal, Bangladesh, India, and Sri Lanka. The welcome news caused me to feel butterflies in my stomach. Laura was excited and delighted by this news as well.

On Wednesday, March 3, I received a phone call from med tech George Benko in Falls Church, Virginia. George offered me a room in his home while I was doing my entry duty studies and examinations. We planned to leave Raul at our home until he finished his commercial marine technology program at Santa Barbara City College. Then he would find tenants and rent out our home in our absence. Selling our custom-built home then was just not one of our options, since we would never be able to afford to buy it back.

My last visit to the *HGE* was on Wednesday, March 17, St. Patrick's Day, with Dr. Flickenger. It was truly the end of an era for me. But Laura and I were anxiously looking forward to leaving on to our next adventure that would most certainly lead to more great stories.

On Saturday, March 27, I was up early as usual and heading for Washington, DC. I was going to be coming back home in June to pick up Laura for our trip to Iran. Just as Vernon Friedell, MD, had been with me on the plane to Washington, DC, before, so were three of my colleagues on this flight. They were going to a medical conference in Washington, DC, as well. We enjoyed each other's company, and there was no fire in the light fixture in the overhead bin on the flight.

Glenn Steele and two other gentlemen greeted me when I landed at Dulles airport. Glenn was to be responsible for keeping me at the right place at the right time and one of the gentlemen, George Benko, was to be my landlord during my overseas training before being sent to Iran. We picked up my luggage, and they drove me to George's home. George had a lovely Cape Cod-style two-story home with a full basement in the lovely wooded suburb of Falls Church. I learned that George had been employed with the government since he was eighteen years old. He had already been a world traveler for years. He would be eligible for retirement before he was fifty years old. We had coffee at George's before Glenn and the other gentlemen left for their homes.

After they left George showed me to my room. It was on the second floor and had a beautiful view, was fully furnished, and had an adjacent bathroom for my use. I received my linens and towels and learned that I would have two other tenants with me. George instructed me on the strict rules that he enforced with the tenants but made me feel welcome.

I got a ride from George that afternoon and got my bearings for the next couple of months. My training would commence at a location that was only four miles from his home.

On Monday, March 29, Glenn picked me up at 7:00 a.m. We stopped with every other car at the security post on the access road just off the highway. The training location was government controlled, and the security personnel were sharp, alert, and observant. They particularly scrutinized my temporary badge as we arrived at the gate. We entered the facility and parked in a restricted area assigned to Glenn. From there it was a quarter-mile walk through the trees to the main office building.

Glenn identified me to the personnel in the main lobby and escorted me after we were cleared. George Benko was already there, and he showed me through the offices. I was introduced to the staff, and they already seemed genuinely pleased that I was coming aboard. I finally came to a nice young receptionist in the office of orientation. I was to report to her for orders for the next several days.

My training there was an intensive week, which consisted of such things as the procedures for computer-generated activity reports on secured lines that were to be linked back to the main offices stateside and to other secured facilities throughout the world. Other training focused on international travel requirements throughout the region I was being assigned to. I learned that, after I reached Tehran, I would be expected to begin setting up itineraries with each of the embassies in my region for my periodic visits. I was to make a visit to five different locations in each two-week period.

Because of the volatility of the subcontinent and Middle East politically at the time, I had classes that taught me physical security measures and precautions for maintaining my personal safety in travel.

Saturday, April 3, and Sunday, April 4, was my first weekend in Washington, DC. I toured around the area and got a flight with a friend of George's over the old Civil War battlefields. I made George and the other tenants dinner on Saturday night, which impressed them.

On Monday, April 5, I drove George's car to Washington National Airport and took a charter flight to Williamsburg, where I was to have my next security training procedures that next week. The training was intensive that week. We were issued military fatigues to replace our street clothes. Meals were served in a mess hall setting. On this day the news reported that Howard Hughes had died. Soon-to-be President Carter announced, "I'm Jimmy Carter, and I'm going to be your next president!" Later in the week I was told that Laura and I would be flying to Tehran on June 11.

On Friday, May 28, my training came to an end. I returned to George's home and began packing out my things and getting myself organized for my departure. On Friday, June 4, I headed for Dulles airport and my flight back to Los Angeles. Randy met me at LAX, and we drove home to a marvelous homecoming.

On Friday, June 11, Laura and I flew to LAX from the Santa Barbara Airport and caught another flight to John F. Kennedy Airport in New York. On June 12, we took off on a Boeing 747 and flew over the Atlantic Ocean and then the south of France. We could see the snowcapped Matterhorn and the Swiss Alps from our window seats. We landed in Rome and stayed two nights there.

On Sunday, June 13, we left Rome and had an unscheduled but interesting stop in Damascus, Syria. The Syrian military security forces were extremely visible. We were instructed to remain on the plane, since we were just picking up passengers bound for Tehran.

We landed at Mehrabad Airport in Tehran at 8:30 p.m. The regional medical officer whom I was replacing met us. He was most cordial and hospitable. In Tehran we had a motor pool that I had been strongly encouraged to use for all my business purposes while there. Laura and I were driven by a limousine with Plexiglas windows and steel shielding inside the doors to our private apartment. Our driver helped us to bring in our luggage and showed us the stocked refrigerator and cupboards. He then gave us a tour of the homemaking amenities and the state-of-the-art security protections of the apartment. I received my orders and was to report to "Det-5" on the following morning. The DOMP mission for me was over, but Laura and I anxiously anticipated our next first overseas adventure.

AFTERWORD

While the project did recover a portion of the *K-129*, a mechanical failure in the grapple caused two thirds of the submarine to break off during recovery. This lost section was to have held many of the most sought items of the mission, including the codebooks and nuclear missiles. Later it was reported that two nuclear-tipped torpedoes and some cryptographic machines were recovered along with six of the *K-129*'s crew. I observed these six crewmen's formal burial at sea, and recently a videotape of the burial was made public.

From March to June 1976, the General Services Administration (GSA) published advertisements inviting businesses to submit proposals to lease the *HGE*. The Lockheed Missile and Space Company submitted a $3-million, two-year lease proposal contingent on the company's ability to secure financing. Unfortunately, Lockheed was unable to secure the financing, and no agency or department of the government wanted to assume the maintenance and operating costs. The *HMB-1* was taken over by the Energy Research and Development Administration (ERDA) and renamed *Otec* (Ocean Technology) to be used for research in ocean thermal-energy conversion.

In September 1976 the GSA turned the *HGE* over to the US Navy for mothballing, and in 1977 she was sent to Suisun Bay in the northeast part of the San Francisco Bay. The *Los Angeles Times* on December 6, 1976, said the project was a total success. At the time, George H. W. Bush was the CIA chief. Unlike William Colby, he absolutely refused to discuss the event.

GLOSSARY OF TERMS

American Academy of Family Physicians (AAFP)
American Medical Association (AMA)
Anti-Submarine Warfare (ASW)
Automatic Station Keeping (ASK)
California Medical Association (CMA)
Capture Vehicle (CV), *Clementine*
Comprehensive Health Examination Clinic (CHEC)
Deep Ocean Mining Project (DOMP)
Department of Occupational and Preventive Medicine (DOPM)
East Coast Trials (ECT)
Energy Research and Development Administration (ERDA)
General Motors (GM)
Delco/General Motors Sea Laboratories Operations (DGMSO)
General Services Administration (GSA)
Hospital/Dispensary (Hospital)
Hughes Glomar Explorer (HGE)
Hughes Marine Barge-1 (HMB-1)
Internal Systems Testing (IST)
Los Angeles International Airport (LAX)
National Association of Underwater Instructors (NAUI)
National Oceanic and Atmospheric Administration (NOAA)
Ocean Technology (Otec)
Program Detail Schedule (PDS)
Project Office (PO), fifth floor of the Hughes Building, Inglewood, California
Santa Barbara County Medical Society (SBCMS)

Santa Barbara Medical Foundation Clinic (Clinic)
Santa Cruz Acoustic Range Facility (SCARF)
Summa Corporation, a subsidiary of the Hughes Tool Company
Target Object (TO), Soviet *K-129* nuclear-armed submarine
Undersea Medical Society (UMS)
United States Naval Station (USNS)

BIBLIOGRAPHY

Varner, Roy, and Wayne Collier. *A Matter of Risk: The Incredible Inside Story of the CIA's Hughes Glomar Explorer Mission to Raise a Russian Submarine.* Ballentine Books, New Ed Edition, 1980.

Sontag, Sherry, Christopher Drew, and Annette Lawerence Drew. *Blind Man's Bluff: The Untold Story of American Espionage.* HarperCollins Publishers, 2000.

Bucher, Lloyd M. *Bucher: My Story.* Doubleday, 1970.

www.ingramcontent.com/pod-product-compliance
Lightning Source LLC
Chambersburg PA
CBHW030926180526
45163CB00002B/480